Myths and Verities in
Protein Folding Theories

Myths and Verities in Protein Folding Theories

Arieh Ben-Naim

The Hebrew University of Jerusalem, Israel

World Scientific

NEW JERSEY · LONDON · SINGAPORE · BEIJING · SHANGHAI · HONG KONG · TAIPEI · CHENNAI · TOKYO

Published by

World Scientific Publishing Co. Pte. Ltd.

5 Toh Tuck Link, Singapore 596224

USA office: 27 Warren Street, Suite 401-402, Hackensack, NJ 07601

UK office: 57 Shelton Street, Covent Garden, London WC2H 9HE

Library of Congress Cataloging-in-Publication Data

Names: Ben-Naim, Arieh, 1934–   author.

Title: Myths and verities in protein folding theories / Arieh Ben-Naim,
    The Hebrew University of Jerusalem, Israel.

Description: New Jersey : World Scientific, 2016. |
    Includes bibliographical references and index.

Identifiers: LCCN 2015031967| ISBN 9789814725989 (hardcover : alk. paper) |
    ISBN 9814725986 (hardcover : alk. paper) | ISBN 9789814725996 (pbk. : alk. paper) |
    ISBN 9814725994 (pbk. : alk. paper)

Subjects: LCSH: Protein folding.

Classification: LCC QP551 .B337 2016 | DDC 572/.633--dc23

LC record available at http://lccn.loc.gov/2015031967

British Library Cataloguing-in-Publication Data

A catalogue record for this book is available from the British Library.

This book is dedicated to the countless students and researchers, who wasted so much time, effort and funds, in a futile search of a solution to the protein folding problem in the wrong direction. And to all new students and researchers who, upon reading this book, will be saved from wasting time, effort and funds. . .

Contents

Preface

Open any book on biochemistry and you are likely to find statements of the following forms:

"It is well known that hydrogen bonds do not contribute significantly to the stability of proteins."
"It is well known that hydrophobic interactions are the most important driving force for protein folding."
"It is well known that the native structure of a protein resides in the global minimum of the Gibbs energy landscape."
"It is well known that the idea of iceberg formation explains the hydrophobic effect."

Such statements are not only typical in textbooks of biochemistry, biophysics and molecular biology, but also frequently appear in the opening sentences of articles in scientific journals.

Usually, such statements will be followed by a few references which presumably contain the "proof" or at least the supporting "evidence" for the validity of the statements. By providing these references the author can continue to build on the opening sentence without feeling the need for further justification, explanation or clarification of the statements.

After all, if something is "well known," why bother with any details which will be at best superfluous.

Most readers of such articles would accept the validity of such statements, especially when the references given are to articles published by famous and well-recognized authorities in the field. However, if you want to check the validity of the statement and look at the articles referred to, you may find similar statements such as "It is well known … ," "It is well established …" or "Convincing evidence was given by …" and more references to earlier literature.

Continuing this process, you will eventually reach the *origin* of the statement, where someone made a *conjecture*, a *hypothesis*, a *plausible* argument or even a *guess* — anything but a *proof* or *convincing evidence* for the validity of the statement.

The purpose of this book is to critically examine the validity of the *original statements* (or hypothesis, conjecture, etc.). All of these statements pertain to the so-called protein folding problem (PFP).

The introductory chapter will briefly present five myths which originated from well-known and well-respected authors. The next five chapters will present in-depth analysis of these five myths and their origins:

(1) From Frank and Evans, "iceberg" idea to "explanation of the hydrophobic effect."
(2) From Schellman's experiments to Fersht's "hydrogen bond inventory argument."
(3) From Kauzmann's conjecture to the "dominance of the hydrophobic effect."
(4) From Levinthal's question to the "solution of Levinthal's paradox."

(5) From Anfinsen's hypothesis to searching for the "folding code," or "predicting" the structure of protein by searching for the global minimum of the Gibbs energy landscape.

Each chapter begins with a short description of a myth, then examines its origin and its evolution, and eventually is debunked.

The last chapter, Chapter 7, is devoted to a few potential candidates for developing into new myths. I am referring, in particular, to two recent "principles" which were introduced into the literature on protein folding. I share with Wedemeyer and Scheraga (2001) the hope that these recent canards will soon vanish from the literature.

All the material discussed in this book has already been published in recent articles in scientific journals. None of these articles were easily accepted for publication. I am well aware of the fact that what I have written has enraged those who propagate these myths. The most surprising finding is that all of these myths continue to thrive and feature even in the most recent textbooks despite the mounting evidence which invalidates them. I will discuss the reasons for such a lingering survival of the unfittest theories in the Epilogue.

Arieh Ben-Naim

Department of Physical Chemistry
The Hebrew University of Jerusalem
Jerusalem, Israel
E-mail: ariehbennaim@gmail.com
URL: ariehbennaim.com

Acknowledgements

I am indebted to the many friends and colleagues who have contributed to my understanding of the protein folding problem. I have enjoyed many discussions with my friend Harry Saroff, to whom I dedicated my previous book, *The Protein Folding Problem and Its Solutions*. I also had many fruitful discussions with Keith Dunker, Guiseppe Graziano, Elisha Haas, Dominik Horinek, Bernard Lavenda, Daumantas Matulis, Sylvia McLain, Roland Riek, Jens Smiatek, Sauro Succi and David Thomas.

In addition, I wish to thank Leonor Cruzeiro, William Eaton, Jeremy England, Walter Englander, Paolo Giaquinta, Daniel Harries, Maik Jocob, Daumantas Matulis, Vladimir Morozov, Roland Netz, David Thomas and Leslie Woodcock for reading parts of or the whole of the manuscript, as well as offering helpful comments and suggestions.

I would like to thank the anonymous reviewer of the article, which I submitted to the *Journal of Chemical Physics* (*JCP*) in May 2013 titled:

"*Myths and Verities in Protein Folding Theories, Part I: Anfinsen's Hypothesis and the Search for the Global Minimum in the Gibbs Energy Landscape*"

I would also like to thank the associate editor Murugappan Muthukumar who accepted the article for publication in *JCP* based on his reading the article and on the opinion of the anonymous reviewer. Unfortunately, this paper was never published in *JCP* (see also the Epilogue).

Finally, I am grateful to my dear wife, Ruby, for her help and patience during the writing of this book.

List of Abbreviations

aa	amino acid
BB	backbone
EL	energy landscape
GEL	gibbs energy landscape
GPF	grand partition function
Hb	hemoglobin
HB	hydrogen bond
HLD	high-local-density
$H\phi I$	hydrophilic
$H\phi O$	hydrophobic
HS	hard sphere
IDPs	intrinsically disordered proteins
KSA	kirkwood superposition approximation
LHS	left hand side
LLD	low-local-density
MI	mutual information
MM	mixture model
PEL	potential energy landscape
PES	potential energy surface

PF   partition function
PFP   protein folding problem
PMF   potential of mean force
RHS   right hand side
SOW   structure of water

1

Introduction: The Protein Folding Problem — Some of the Myths Associated with the Protein Folding Theories

During the years 2012 and 2013, while I was writing my book *The Protein Folding Problem and Its Solutions*, I asked some of my friends and colleagues what they thought the protein folding problem (PFP) is. The majority of the answers are summarized as follows:

(A) The stability problem: What are the factors responsible for the stability of the native 3D structure of the protein?
(B) The kinetic problem: What are the factors that "guide and speed" the folding of the protein?
(C) The folding code: Can we *predict* the 3D structure of a protein given the sequence of amino acids?

Having received these answers, I asked the next question: "Which, in your opinion, is the most important (challenging, difficult, daunting, etc.) of the three subproblems listed above?" To my surprise, most pointed to the third one.

In fact, one of the answers was: "You solve the protein folding problem only when I give you a sequence of amino acids, and you should be able to give in return a 3D structure." This kind of formulation of the PFP is very common in the literature. Here is an example. In a recent review article by Orevi *et al.* (2014), we find the following opening sentence:

> The protein folding problem would be considered "solved" when it will be possible to "read genes," *i.e.* to predict the native fold of proteins, their dynamics and mechanism of fast folding based solely on sequence data.

Note that in this statement the authors claim that "reading" the sequence (of either the amino acids or the corresponding bases on the DNA) should provide not only the *structure* of the native protein but also the dynamics and mechanism of the folding process. As I will argue in Chapter 6, such statements are likely exaggerated wishful thinking. One cannot expect to predict the *structure* "based solely on sequence data," let alone predict the "dynamics and mechanism."

Having received this information, I asked the last question: "Suppose the folding code is the main problem — What kind of answer would you accept as a valid solution to this problem?" The possible answers I suggested were:

(1) You give me a sequence of amino acids. I will synthesize the polypeptide in a physiological solution. If it folds, I will determine its structure experimentally, and send it back to you as an answer.

(2) You give me a sequence of amino acids. I will do a simulated experiment on a computer, and if it folds into a stable 3D structure, I will give it to you as an answer.

(3) You give me a sequence of amino acids. I will *read* it through. I will not do any experiment, or any simulated experiment on a computer. Just from reading the sequence, I will be able to translate the sequence into a 3D structure.

For this question, most people pointed to answer #2 as the acceptable answer to the PFP.

For a long time, I believed the PFP to consist of the three parts A, B and C mentioned above. While it was relatively easy to understand the first two problems, I never understood what people really meant by a "folding code." I asked many experts in the field, but all I could get was some statement like "You give me a sequence and I will give you a 3D structure." (See quotation above). That is certainly not a code in the usual sense of the term. Therefore, I have concluded that question C has no general conclusion [Ben-Naim, (2013)]. What remained of the PFP were questions A and B. These are no easy questions. Unfortunately, most of the answers given to these questions involved the idea of the hydrophobic effect.

Recently, it has become clear that the hydrophobic effect is not only ill-defined and ill-explained in itself, but also it cannot explain the stability problem (A) or the kinetic problem (B). Instead, the more powerful hydrophilic effects can answer both of the problems A and B. In this sense, I maintain that the various hydrophilic effects provide answers to the two parts of the PFP which are *answerable*. The "folding code" part, considered by many to be the main part of the PFP, has no general solution and can be removed from the PFP.

There are five main myths that will be discussed in detail in the next five chapters. All of them are associated with the PFP and, in my opinion, clinging to these myths is the main reason why the PFP is still considered to be one of the most challenging unsolved problems in molecular biology.

The five myths discussed in the following chapters are arranged according to the chronological order of their origins. This order is not necessarily the same as their relative importance. Each of them had not only caused great confusion in the field, but also contributed to the lingering myth, and the continuous search for solutions in the wrong directions. In particular, the search for a folding code was not only in the wrong direction, but it was a search for a code that does not exist.

In the remaining part of this chapter, I will survey some of the ideas that were involved in the PFP.

"In the beginning," hydrogen bonds (HBs) reigned supreme in the field of proteins. As soon as biochemists realized the importance of HBs for understanding the properties of liquid water, and as soon as the main features of the secondary structure of proteins such as helices and β-sheets were discovered, it was only natural to conclude that HBs are the main factor that contributes to the stability of the structure of the protein. The main proponent of this idea was Linus Pauling. One can follow the "evolution" of this idea by comparing the successive editions of Pauling's book *The Nature of the Chemical Bond* [Pauling (1939, 1948, 1960)]. No doubt Pauling believed that HBs are the major factor that contributes to the stability of the structure of both proteins and nucleic acids. These views prevailed until about the mid-1950s.

In 1955, John Schellman published a series of articles in which he analyzed the association between two urea molecules in water. Each urea molecule (Figure 1.1), has two functional groups, amine and carbonyl; one can serve as a donor and the other as an acceptor for hydrogen bonding (Figure 1.2). In fact, there are two possible dimers between urea molecules involving one or two HBs, Figure (1.3).

Following Schellman's analysis of the dimerization of urea molecules in aqueous solutions, people concluded that hydrogen bonding (either *inter-* or intramolecular) could

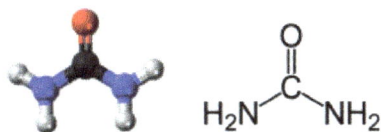

Fig. 1.1. Urea molecule having both a donor and an acceptor for hydrogen bonding.

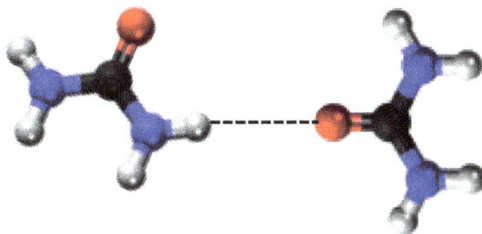

Fig. 1.2. A hydrogen bond formed between the amine and the carbonyl groups of urea.

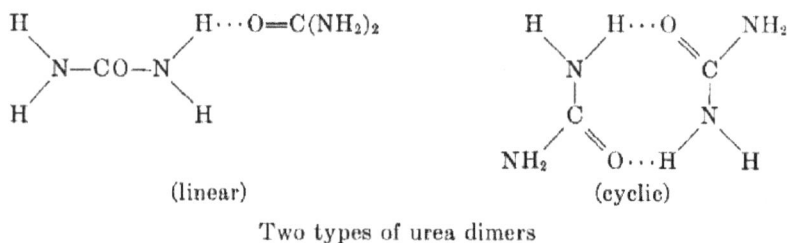

(linear)   (cyclic)

Two types of urea dimers

Fig. 1.3. Two types of urea dimers. From Kauzmann (1959).

not contribute significantly to the driving force for the dimerization of urea and, by generalization, also for protein stability.

Although Schellman himself denied that he reached this conclusion, he did make a serious error in the very writing of the following stoichiometric reaction Figure 1.4.

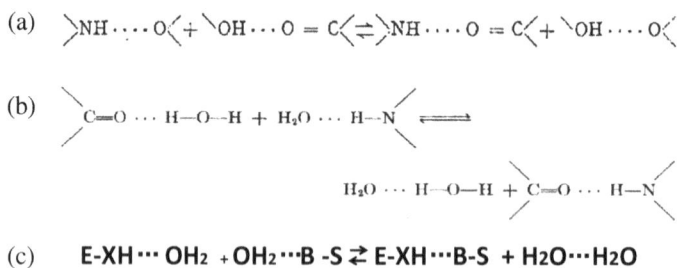

Fig. 1.4. The stoichiometric reaction as written by: (a) Schellman (1955), (b) Kauzmann (1959), and (c) Fersht (1999).

Just looking at this equation, one can conclude that when an HB is formed between two urea molecules in water (or by any two groups in a protein), two HBs are "lost" and two HBs are "gained" in the "reaction." Hence, whatever the strength of the HB energy is, one cannot expect a significant contribution from the HB to the stability of proteins.

This erroneous conclusion has been quoted even to this day in many textbooks of biochemistry. It was referred to as the "HB inventory argument" by Fersht. We will discuss in more detail the evolution of this erroneous conclusion in Chapter 3. Here, we will continue with the historical development of the ideas concerning the PFP.

In 1959, Kauzmann wrote a very influential review article on "Some Factors in the Interpretation of Protein Denaturation." This article is famous for its introduction of the idea of the hydrophobic bond into the PFP.

It seems to me that not too many scientists realized that the inception of the idea of the $H\phi O$ effect in proteins followed the conclusions based on Schellman's experiments. To put it more graphically, the enormous edifice of the *hydrophobic effect* was built on the ruins of the hydrogen bonding roles in protein stability.

Following Schellman's experiments, Kauzmann (1959) concluded:

> *Hydrogen bonds, taken by themselves, give a marginal stability to ordered structures.*

"Ordered structures," in the context of Kauzmann's article, meant the 3D structure of the native protein.

Recall that until 1955, HBs were considered the *most important* factor in the stability of the protein. Now, Schellman's experiments seem to indicate that HBs are not important. This fact has left a kind of "vacuum" in the explanation of the stability of the structure of the protein. If HBs are not important, then *what* is?

Kauzmann came up with a brilliant idea. It was known that the solubility of nonpolar molecules in water is extremely low (see also Chapters 1 and 3). This fact may be translated in terms of Gibbs energy changes. Low solubility means large *positive solvation* Gibbs energy. Transferring a solute from water to an organic liquid involves a large *negative* change in Gibbs energy.

Kauzmann noticed that in the process of folding of a protein, many nonpolar groups which are initially exposed to water are found in the interior of the 3D structure of the folded protein. (See Chapter 4, Figure 4.2).

His idea was simple and, at that time, very convincing. The transfer of a nonpolar *group* (say, methyl, isopropyl, etc.) from water into the interior of the protein is very similar to the transfer of a nonpolar *molecule* from water into an organic liquid (say, alcohol, benzene, hexane, etc.). Figure 4.2 of Chapter 4 shows the two processes. In both, a nonpolar group is transferred from an aqueous to a non-aqueous environment. This suggests that the Gibbs energy of transferring a nonpolar group from being exposed to water into the interior of the protein might be large and negative.

A typical protein of, say, 150 amino acids might have about 30% nonpolar (or hydrophobic) groups. If each of these groups, when transferred from water to the interior of the protein, contributes between -2 and -3 kcal/mol, then we can expect a very large negative contribution to the stability of proteins due to this effect — which at that time was referred to as the hydrophobic bond.

This idea captured the imagination of most, if not all, of those who were puzzled by the stability of the proteins. Almost no one questioned the validity of this idea. I myself was fascinated by Kauzmann's idea, and I spent almost 15 years working on the so-called hydrophobic effect.

In 1980, I summarized in a monograph [Ben-Naim (1980)] what was known at that time about the hydrophobic effects. I have made a clear distinction between hydrophobic solvation and hydrophobic interaction. In the preface to the book, I wrote:

> *However, in spite of my researches in this field over almost ten years, I cannot confirm that there is at present either theoretical or experimental evidence that unequivocally demonstrates the relative importance of hydrophobic interactions (HI$_s$) over other types of interactions in aqueous solutions.*

I have discussed my doubts with Kauzmann, and he certainly agreed with me that there was no hard evidence that the various hydrophobic effects are really the most important factors in protein stability.

About ten years later, while I was spending a sabbatical year at the NIH, I examined the entire question of *all* possible solvent-induced effects on the stability of the protein.

Two immediate conclusions from this study were quite stunning to me. First, the so-called HB inventory argument is fundamentally wrong. Second, the Kauzmann model for the transfer of a nonpolar group into the interior of the protein was overexaggerated. More about these two conclusions is discussed in Chapters 3 and 4. In the late 1980s, I summarized my findings in a few articles [Ben-Naim (1989,

1990a, 1990b, 1991a, 1991b)] and in book [Ben-Naim (1992)].

The third outcome of that study was the discovery of new hydrophilic ($H\phi I$) effects which are different from direct hydrogen bonding, and in some cases stronger than direct HBs.

Incredibly, despite the proof of the invalidity of the HB inventory argument, despite the evidence of the inadequacy of the Kauzmann model for the $H\phi O$ effect, and despite the overwhelming evidence that various $H\phi I$ effects are much stronger than the corresponding $H\phi O$ effects, most biochemistry books continue to propagate the erroneous ideas about the insignificance of HBs, and the importance of the $H\phi O$ effect to the stability of the protein. These two myths are discussed in Chapters 3 and 4. The myths continue to linger in spite of all the evidence debunking them.

In the late 1960s, people also asked questions regarding the kinetics of the process of protein folding. A landmark article by Levinthal (1968) articulated the problem.

An unfolded (or denatured) protein has many rotational degrees of freedom. Levinthal asked what the factors are that *speed* and *guide* the protein on its folding pathway from the unfolded to the folded 3D structure of the native protein.

Levinthal argued that if the protein would have searched the configurational space at random, it would have taken eons to reach the stable 3D structure of the protein. Thus, there must be some factors that *guide* and speed the folding of the protein. Later, this argument came to be known as the *Levinthal paradox*. Many scientists speculated about this paradox and suggested solutions to it. Unfortunately, there is no paradox at all, and Levinthal never saw a paradox in his estimates. Nevertheless, the myth of the Levinthal paradox still lingers in the biochemical literature. We will devote Chapter 5 to the kinetic aspect of protein folding. We will also see that *hydrophilic forces* are probably the best candidate for answering the question raised by Levinthal.

Having dealt with some of the most outstanding myths involving the hydrophobic effects, we will turn to discussing two other myths (in Chapter 5) which were spawned from Anfinsen's thermodynamic hypothesis. These two myths have nothing to do with the $H\phi O$ effect, and yet the biomedical literature is replete with them. One is the existence of a folding code, and the other is the belief that the structure of a protein resides in the global minimum of the Gibbs energy landscape. The latter is often presented as a "well-known" fact.

While I cannot offer a proof that a folding code does not exist, I can provide some compelling evidence that such a code is unlikely to exist. Similarly, one cannot prove that the native structure of a protein does not reside in the global minimum of the Gibbs energy landscape. However, it is possible to show that this belief is a result of misinterpreting Anfinsen's hypothesis or, better yet, of misunderstanding the second law of thermodynamics. In other words, it does not follow from the second law that the native structure of the protein necessarily resides in the global minimum of the

Gibbs energy landscape. Both of these myths are discussed in detail in Chapter 6.

Finally, in Chapter 7, I have made some notes regarding some concepts that have appeared recently in the literature on protein folding, which in my opinion have the potential to evolve into new myths.

2

From Frank and Evans' Iceberg Formation Conjecture to the Explanation of the Hydrophobic Effect

2.1 Abstract

In 1945, Frank and Evans published an article on the thermodynamics of solvation of inert solutes in water and in other liquids. They found that the entropy of solvation of these solutes is much larger and negative in water as compared with the entropy of solvation of the same solutes in other liquids. In order to explain these findings, the authors *conjectured* that when an inert solute dissolves in water it forms some kind of "icebergs" around it. This idea has captured the imagination of many scientists for more than half a century. How can an inert solute, weakly interacting with water molecules, form an "iceberg"? The contrast between the "innocent," small atom of argon or neon and the "magnificent" structured icebergs must have impressed many scientists, including me. This was also the reason I chose to study the "thermodynamics of solvation of inert gases in aqueous solutions" in my PhD thesis.

Frank and Evans did not *prove* that an inert solute builds up icebergs around it; nor did they provide any explanation as to why an inert solute should form icebergs. All they did was to interpret the negative change in entropy in terms of increasing the *order*, or equivalently increasing the *structure* of water. Yet, this idea was used by many scientists to explain the entropy and the enthalpy of solvation of the nonpolar solute in water. In addition, it was used by many in explaining the hydrophobic effect. We show here that the idea of iceberg formation, or a variation of this idea, can explain the outstanding large and negative entropy and enthalpy of solvation of inert solutes in water, provided that one first explains how a simple solute forms icebergs. However, this idea cannot explain the Gibbs energy of solvation of such solutes, or the hydrophobic effect. Furthermore, the iceberg formation idea cannot be used to explain the large and positive partial molar heat capacity of nonpolar solutes in water.

2.2 The Origin of the Myth and Some Historical Notes

In 1945, Frank and Evans published an important and very influential article. In this article they summarized what was known at that time about the outstanding thermodynamic quantities of solvation of simple nonpolar solutes in water.

Table 2.1 provides some experimental data on the solubility of argon in water and in some other liquids. Note that the solubility of argon is about an order of magnitude smaller in

Table 2.1   Solubility of Argon (in Terms of Ostwald Absorption Coefficient γ) in Water and in Some Organic Liquids at Two Temperatures [Ben-Naim (2009)]

Solvent	15°C	25°C
Benzene	0.232	0.240
Cyclohexane	0.330	0.334
n-hexane	0.474	0.472
n-heptane	0.411	0.415
n-octane	0.355	0.367
Fluorobenzene	0.291	0.298
Chlorobenzene	0.202	0.204
Bromobenzene	0.153	0.157
Iodobenzene	0.104	0.109
Toluene	0.240	0.249
Nitrobenzene	0.100	0.105
Water	0.0396	**0.0341**

water than in the other solvents. Also note that the solubility of argon in water increases with temperature, whereas in other solvents it changes very little.

Table 2.2 shows the solvation Gibbs energy, entropy and enthalpy of methane in water and in some organic liquids. Figure 2.1 shows the Gibbs energy of solvation of xenon in water and in some linear alcohols. Figure 2.2 shows the entropy and enthalpy solvation of xenon in water and in some linear alcohols. Figure 2.3 shows the change in the Gibbs energy of solvation of methane in mixtures of water and ethanol.

As can be seen from Table 2.2 and Figures 2.1 and 2.2, the Gibbs energies of solvation of the nonpolar solutes in

Table 2.2   Values of the Solvation Gibbs Energy, Entropy, Enthalpy, and Partial Molar Heat Capacity of Methane in Water and in Some Nonaqueous Solvents at Two Temperatures [Ben-Naim (2009)]

Solvent	t (°C)	ΔG_S^* (cal (mol^{-1})	ΔS_S^* (cal mol^{-1}K^{-1})	ΔH_S^* (cal (mol^{-1})	$\Delta C_{P,S}^*$ (cal mol^{-1}K^{-1})
Water	10	1747	−18.3	−3430	53
	25	2000	−15.5	−2610	
Heavy water	10	1703	−19.2	−3740	52
(D$_2$O)	25	1971	−16.5	−2940	
Methanol	10	343	−2.6	−390	−21
	25	390	−3.7	−710	
Ethanol	10	330	−3.2	−570	−5
	25	380	−3.5	−650	
1-propanol	10	345	−4.3	−880	25
	25	400	−3.0	−500	
1-butanol	10	369	−2.8	−420	−33
	25	430	−4.5	−910	
1-pentanol	10	399	−3.3	−530	−7
	25	450	−3.6	−630	
1,4-dioxane	10	538	−0.8	+310	−6
	25	553	−1.1	+220	
Cyclohexane	10	154	−1.9	−390	11
	25	179	−1.4	−230	

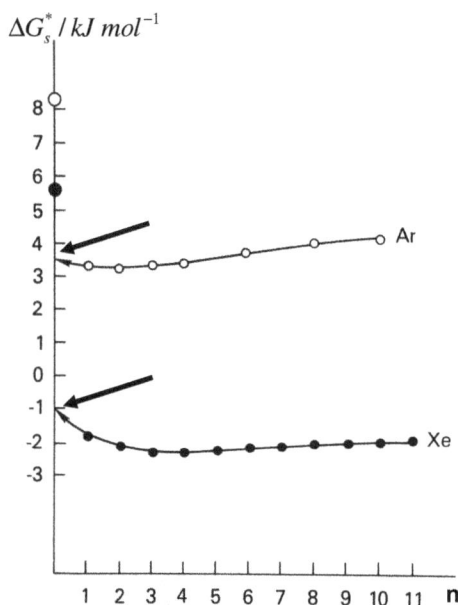

Fig. 2.1. Solvation Gibbs energy of argon (open circles) and xenon (full circles in a series of linear alcohols). N is the number of carbon atoms in the alcohol. The arrows indicate the extrapolated values for $n = 0$. The experimental values for water are shown as open and full circles at $n = 0$.

water are positive and much larger than in other liquids. On the other hand, the entropies and enthalpies of solvation in water are large and negative in water as compared with the nonaqueous solvents.

It should be noted that the definition of the solvation process discussed in the paper of Franks and Evans is different from the definition we use in this book. In this book, we will use the definition of the solvation *process* as suggested by Ben-Naim (1980, 2006). The process is the transfer of a molecule from a fixed position in an ideal gas phase to a fixed position in the liquid (Figure 2.4). More details can

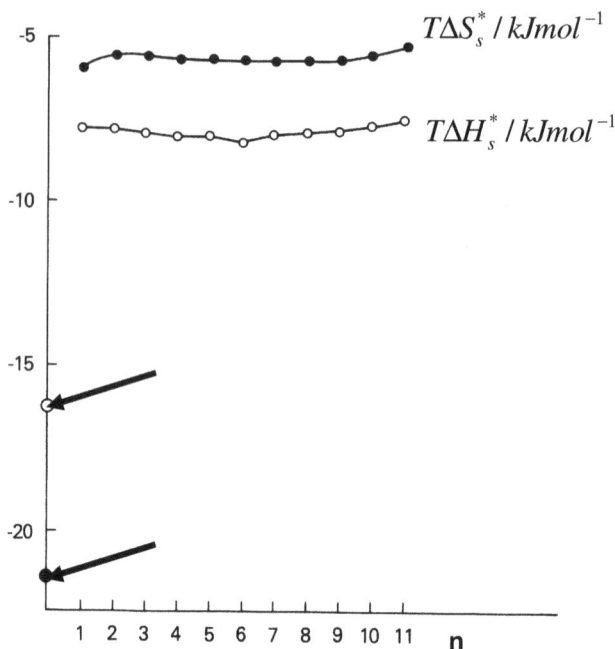

Fig. 2.2. Solvation enthalpy (open circles) and entropy times T (full circles) for xenon in a series of linear alcohols. n is the number of carbon atoms in the alcohol. The experimental values for water are shown as open and full circles at $n = 0$. All values re for 1 Atm and 20°C.

be found in Appendix A. The difference in the definition of the solvation process changes the *magnitude* of the solvation thermodynamic quantities. However, this difference does not affect the conclusions reached in this chapter.

Looking at the values of ΔS_s^* in Table 2.2 reveals that the solvation entropy of methane in water (and in D_2O) is much larger (in absolute magnitude) than in any other organic liquid. In order to understand these values, let us consider first the solvation of KCl in water.

Figure 2.5 shows schematically the arrangement of water molecules around two ions, K^+ and Cl^-. It is known that the

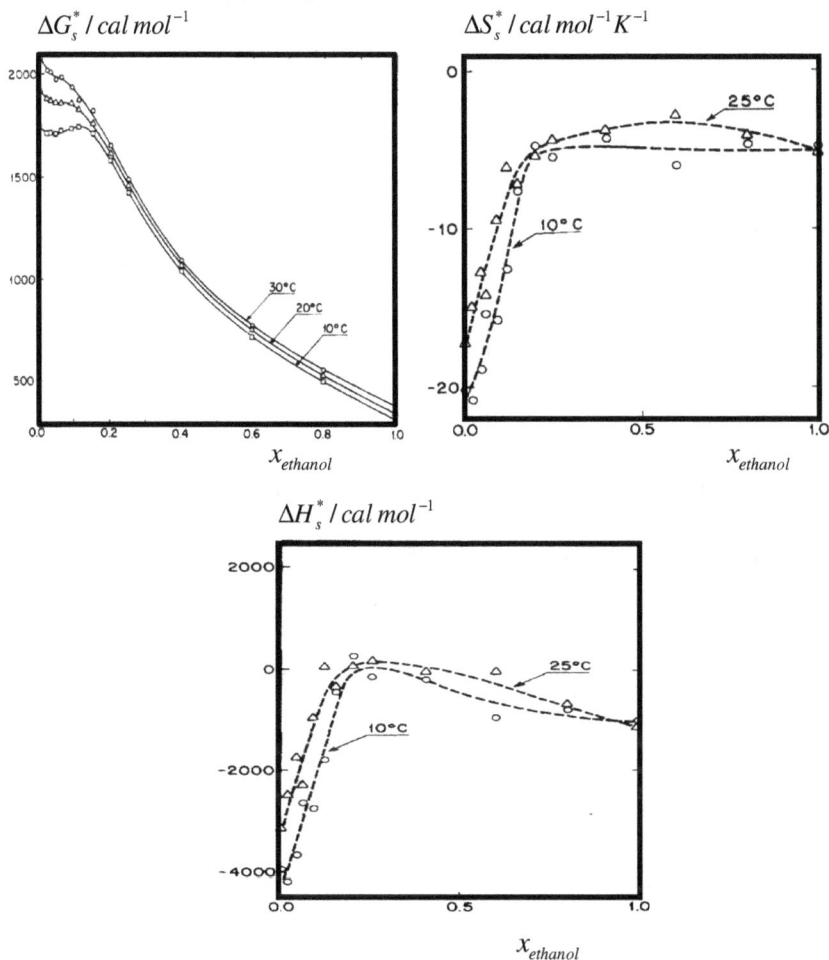

Fig. 2.3. The solvation Gibbs energy, entropy and enthalpy of methane in mixtures of water and ethanol at three temperatures.

solvation entropy of KCl in water is large and negative. Now, consider the following "reaction":

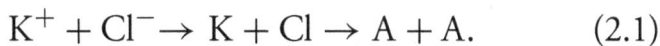

$$K^+ + Cl^- \rightarrow K + Cl \rightarrow A + A. \qquad (2.1)$$

In this process, we first transfer the electron from the chlorine ion to the potassium ion. This will produce two

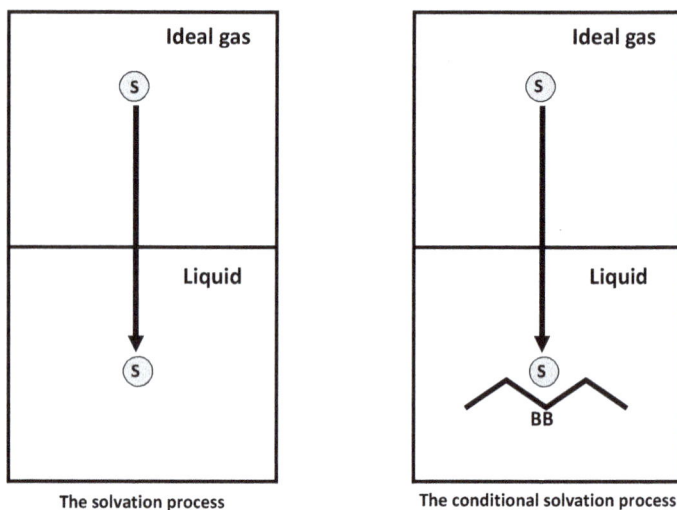

Fig. 2.4. The process of solvation and conditional solvation.

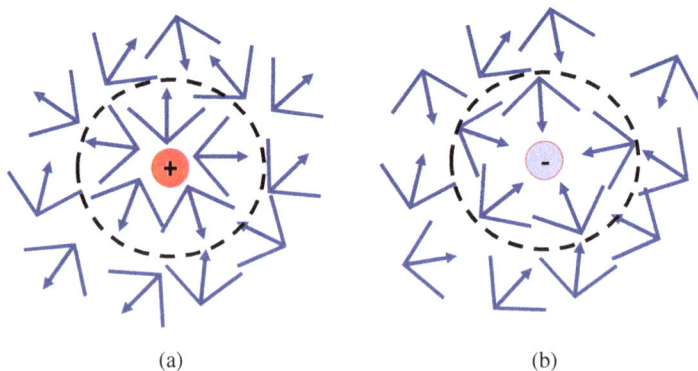

Fig. 2.5. Arrangement of water molecules around (a) a positive and (b) a negative ions.

neutralized atoms, K and Cl. Next, we replace K and Cl by two argon atoms, A.

It is easy to understand why the entropy of solvation of KCl in water is large and negative. The electric field around the charged ion is very strong. This field will force the dipole

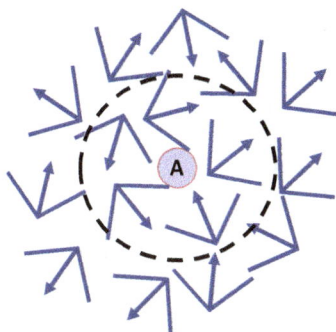

Fig. 2.6. A random arrangement of water molecules around an argon atom.

moments of the water molecule to orient themselves as shown in Figure 2.5. Accepting for the moment the interpretation of entropy as a measure of order, we can associate the negative solvation entropy of KCl with this preferential orientation, or ordering of water molecules around the ions.

Once we remove the charges on the ions, we also "turn off" the electric field around the ions (Figure 2.6). Therefore, we cannot expect any ordering effect around the ions, and hence we expect that the entropy of solvation of the two neutral atoms will be smaller (in absolute values) than for KCl. Equivalently, we should expect that in the "reaction (2.1)", the entropy change will be positive.

The experimental fact is that the entropy of solvation of two argon atoms (roughly having the same size as the neutral atoms K and Cl) is even *more* negative than the entropy of solvation of KCl.

These findings are quite puzzling. Why should the entropy of solvation of two neutral atoms be more negative than that of the charged atoms?

The large negative entropy of solvation of a simple inert solute was quite striking and difficult to explain. This fact was even more surprising when it was compared with the entropy of solvation of two ions such as K^+ and Cl^-, which have roughly the same size as argon atoms.

The main problem is how to explain these outstanding values of the entropy and the enthalpy of solvation of the inert solutes in water on a molecular level.

Two theories were suggested. The first was by Eley (1939, 1944), and it was based on a lattice theory of water (a very common approach to liquids at that time). Eley assumed that there are a *fixed number* of holes, or cavities in liquid water. When an inert gas dissolves in water, it is confined to such fixed numbers of holes. In this view, the reduction in the translational entropy of the molecules in the process of solution explains the negative entropy change. A completely different picture was suggested by Frank and Evans (1945). Those authors did not assume any fixed number of holes, or a fixed *structure* of the water. Instead, they proposed that the dissolved molecules create some kind of a new structure in water, referred to as "icebergs." In their words:

> When a rare gas atom or non-polar molecule dissolves in water at room temperature, it modifies the water structure in the direction of greater "crystallinity" — the water so to speak builds a microscopic iceberg around it.

It should be noted that at the time of the publication of Frank and Evans' article the concept of *entropy* was understood as a measure of order or disorder. The negative entropy of solution of nonpolar molecules was interpreted as an increase in *order*, or equivalently as an increase in the *structure* of water. Frank and Evans did not *explain* why an inert solute would cause an increase in the structure. They simply translated the experimental fact of a negative change in entropy into the language of increasing the *order*, or the *structure* of the system. For the present discussion we accept the "order–disorder" interpretation of the entropy, although this interpretation cannot be justified [Ben-Naim (2008, 2012d)], and in fact does not hold in general.

Both theories tried to explain the negative entropy change in terms of increase in order. Confining solute molecules to a fixed number of holes is clearly conceived as a more "ordered" state, compared with that of molecules in the gaseous phase. The smaller the number of holes, the more "localized" the gas molecules, and hence the entropy of solvation is expected to be negative. On the other hand, Frank and Evans' picture relegates the ordering effect to the water, i.e. building up new structures.

Eley's theory did not survive long. The idea of water having a fixed number of holes was untenable. On the other hand, Frank and Evans' ideas have survived until today. It survived not because it was proven to be correct, but rather because it was a nice pictorial idea that captured the imagination of most scientists working in the field of aqueous solutions [Tanford and Reynolds, (2001)].

Not only did the iceberg idea survive, but it was used over the years to explain many other phenomena associated with aqueous solutions. It was much later that the structuring of water, by an inert solute, was reformulated (Ben-Naim, 1974) not in terms of building up a *new structure* (as in the case of ionic solutes, or what Frank and Evans envisaged by icebergs), but in terms of *enhancement* of the already existing structure of liquid water. This was only a shift in formulating the problem. It was not yet an answer to the question of why an inert solute would enhance the structure of water. It was only much later that a firm argument, based on the Kirkwood–Buff theory, was provided [Ben-Naim (1992, 2009)].

2.3 The Evolution of the Myth

In many articles, as well as in many biochemistry textbooks, one finds statements alluding to an "explanation" of the hydrophobic effect by invoking the idea of iceberg formation. We present here only a few examples taken from the recent literature. Underlying these statements are the following arguments:

(i) The solvation entropy dominates the solvation Gibbs energy. This means that in the equation

$$\Delta G_s^* = \Delta H_s^* - T \Delta S_s^*,$$

the quantity $|T \Delta S_s^*|$ is much larger than $|\Delta H_s^*|$.

(ii) The enhancement of the structure of water (by iceberg formation or any equivalent solvent-induced effect) explains the large negative entropy of solvation.

(iii) Therefore, it follows from (i) and (ii) that the enhancement of the structure which is responsible for the large positive $- T \Delta S_s^*$ is also responsible for the large positive ΔG_s^*.

This argument is repeated very often in the literature [Tanford and Reynolds (2001); Snyder *et al.* (2013)]. It sounds very logical: if structure enhancement is responsible for the large negative value of $T\Delta S_s^*$, and if this term is dominant in ΔG_s^*, then it follows that the structural enhancement also explains the large positive ΔG_s^*, and hence explains the hydrophobic effect.

This argument, though plausible, is incorrect. We will discuss this in the succeeding sections of this chapter.

In his influential review, Kauzmann (1959) writes:

> *Frank and Evans concluded that when a non-polar molecule is present in water, the water molecules in the immediate vicinity must arrange themselves into a quasi-crystalline structure in which there is less randomness and somewhat better hydrogen bonding than in ordinary liquid water at the same temperature. They called these structures "icebergs," although they were careful to point out that the structure present in the icebergs need not be the same as that in ordinary ice, and that different structures might be found in the icebergs of different systems.*

It should be noted that Frank and Evans were indeed careful to leave the structure of the "icebergs" unspecified. However, Kauzmann's comment that *the structure present in the icebergs need not be the same as that in ordinary ice* is somewhat ironic. As we will see later in this chapter, in order to get a significant structural change in the solvent, the "structure of the iceberg" must be the *same* as the structure which is already present in liquid water.

Kauzmann (1993), in his article "Reminiscences from a Life in Protein Physical Chemistry," writes:

> *It was, however, the paper of Frank and Evans (1945) that most clearly set the stage for what I believe is the correct picture of the hydrophobic bond....*

... Frank and Evans ... found ... that for the transfer into water ... there is a considerably large entropy decrease. Frank and Evans concluded that this was caused by the formation of what they called "icebergs" (which can be considered a synonym for the term "structure") in the liquid around the solute molecules.

Similarly, Tanford and Reynolds (2001) in their review of the hydrophobic effect mention the idea of iceberg formation:

The pictorial interpretation, crude as it was, captured the imagination of protein chemists — it provided evidence for the disproportionate strength of hydrogen bonds in water, and engendered confidence in the hydrophobic concept.

On the same subject, Wu and Prauznitz (2008) write:

It is well documented that hydrophobicity of a solute in water is not because of its lack of attraction with water molecules but because of the entropy penalty in reorganization of water molecules, to preserve a highly organized hydrogen-bonding network.

They conclude the article with:

Although our conclusions may not be directly applicable to protein folding ... we expect that the free energy of solvation and the hydrophobic interaction discussed here are relevant to the properties of hydrophobic groups at the surface of biological molecules.

Note that the term "hydrophobicity" is ambiguous. There are at least two different hydrophobic effects. One is the solvation Gibbs energy of a $H\phi O$ solute and the second is the potential of mean force (PMF) between two $H\phi O$ molecules in water. For more details, see Ben-Naim (1980, 2011).

More recently, Snyder *et al.* (2013) write:

One of the oldest — and now most pervasive — rationales for a single hydrophobic effect is the formation of "structured" or "iceberg-like" water near polar solutes, as proposed by Frank and elaborated by Kauzmann, Tanford and others. This model rationalizes the transfer of small, simple hydrophobic molecules such as methane or ethane from non-polar phase… to an aqueous phase: the free energies of these transfers at room temperature, are unfavorable and seem to be dominated by a large, unfavorable entropic term. The iceberg model postulates the unfavorable entropy of transfer results from a network of structured water that forms around the hydrophobic molecules.

All the above interpretations of the hydrophobic effect in terms of "iceberg" formation are basically fallacious. As we will see in this chapter, it is true that $|T\Delta S_s^*|$ of nonpolar solutes is larger than the absolute value of the enthalpy of solvation. It is also reasonable to explain the large negative entropy solvation of nonpolar solutes in terms of enhancement of the structure of water. From these two facts one cannot conclude that the iceberg formation is responsible for the large positive Gibbs energy of solvation, or the PMF of two $H\phi O$ molecules in aqueous solutions.

2.4 Some Theoretical Questions

In the following sections of this chapter, we will raise a few theoretical questions and provide some answers. More specifically:

(i) Why would a simple nonpolar solute form icebergs?

(ii) How does iceberg formation explain the entropy and the enthalpy of solvation?

(iii) Can iceberg formation explain the hydrophobic effects (either solvation or PMF)?

(iv) Can iceberg formation explain the large positive partial molar heat capacity of such solutes in water? (Sometimes this is referred to as the "hydrophobic signature.")

2.4.1 *How Can a Simple Nonpolar Solute, Such as Argon, Form "Icebergs"?*

The idea of iceberg formation was speculated on by Frank and Evans at a time when very little was known about the theory of water and aqueous solutions. Frank and Evans were careful enough to leave the description of the structure of the icebergs unspecified. Whatever they meant by the *structure* of the icebergs, they left the main question posed in the title of this subsection unanswered.

As will be clear from the following discussion, it is almost impossible to argue that a simple nonpolar solute will form some kind of a *new structure* in water that does not exist in the pure liquid. On the other hand, it is more plausible to argue that a simple solute could enhance the "degree of structure" that already exists in water. These two structural effects are shown schematically in Figure 2.7. In this figure, the term "iceberg" is used in the sense of a *new structure* which either does not exist in pure water or exists but in an extremely low concentration (or having a low probability of occurrence). The distinction between these two structural effects is important. If the iceberg structure does not exist in pure water, or has a negligible probability of occurrence, then one cannot argue that an inert solute will form or enhance such a structure. On the other hand, if we assume that water already has a high degree of structure (see below for the definition), then one can argue in favor of enhancement of such a structure. Even in this new formulation of the enhanced structure of water, the question of how the solute does that is far from trivial. To do this, one must first have a clear definition of the structure of water.

(a)

N_L^0, N_H^0 N_L^1, N_H^1

(b)

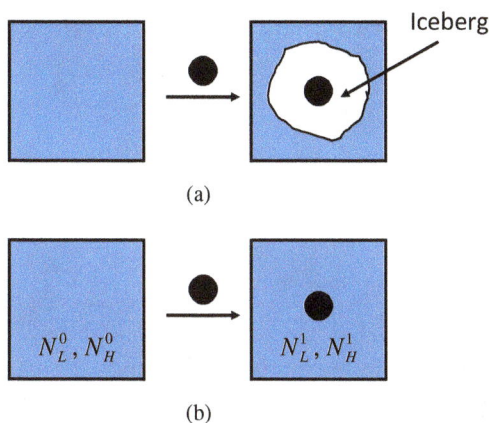

Fig. 2.7. Two different structural effects induced by a solute. (a) Formation of a new structure, referred to as "iceberg." (b) Enhancement of existing structure.

Instead of a solute forming a new structure (as is the case for ionic solutes which form new "structure," or new "order" which does not exist in pure water), it was suggested that the solute enhances the degree of structure which already exists in pure liquid water. In its simplest form (which can be made exact; see also Appendix B) we can *define* the structure of water in terms of *species* (or quasicomponents) which we perceive as being more structured or less structured.

Let L and H be defined as the low-local-density (LLD) water molecules and high-local-density (HLD) water molecules (Figure 2.8). The question of changes in structure induced by the nonpolar group is now formulated in terms of the Le Chatelier principle, which is written as (Appendix C)

$$\left(\frac{\partial N_L}{\partial N_s}\right)_{eq} = -(\mu_{LL} - 2\mu_{LH} + \mu_{HH})^{-1} \left(\frac{\partial \Delta\mu}{\partial N_s}\right)_*,$$

$$(2.2)$$

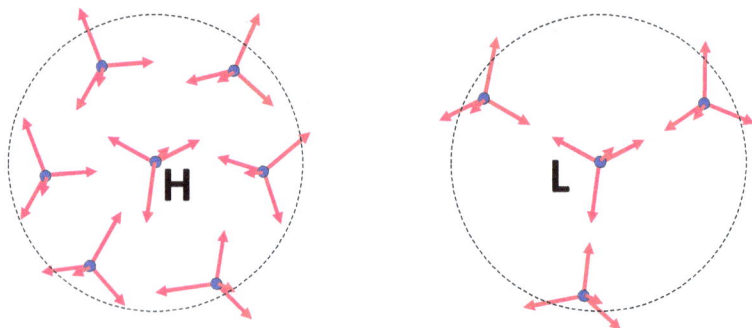

Fig. 2.8. Water molecules having high-local-density (HLD) and low-local-density (LLD).

where N_s is the number of solute molecules s, N_L is the average number of LLD water molecules $\Delta\mu = \mu_L - \mu_H$ (μ_L and μ_H are the chemical potential of L and H, respectively) and $\mu_{\alpha\beta} = \frac{\partial^2 G}{\partial N_\alpha \partial N_\beta}$ is the second derivative of the Gibbs energy. The quantity $\mu_{LL} - 2\mu_{LH} + \mu_{HH}$ must be positive at equilibrium (see also Appendix D).

The derivative on the right hand side (RHS) of (2.2), denoted $*$ is at constant N_L and N_H. On the other hand, the derivative on the left hand side (LHS), denoted "eq," is along the equilibrium line for the conversion reaction between L and H. Equation (2.2) is very general. It applies to any classification of water molecules into two species. It should be noted that such a mixture model approach to the theory of liquid water can be made exact (see Appendix B).

The meaning of Eq. (2.2) can be best explained in terms of a two-step process, as depicted in Figure 2.9. We start with pure water at some temperature T and pressure P. The water molecules are classified as LLD molecules denoted L, and HLD molecules denoted H. The average number of each of

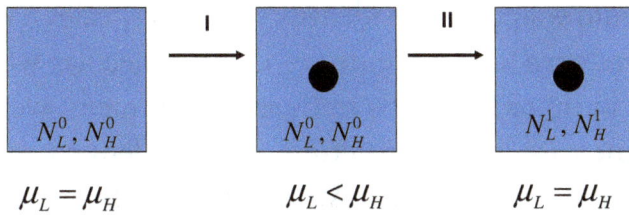

Fig. 2.9. Addition of a solute molecule in two steps: (I) Freezing-in the equilibrium between L and H, and adding the solute, then, (II) relaxing the constraint on fixed numbers of L and h molecules.

these species is denoted N_L^0 and N_H^0, respectively, and the total number of water molecules is $N_w = N_L^0 + N_H^0$. We note here that there are many ways of defining the H and the L species, based on the generalized molecular distribution [Ben-Naim (1974)].

We now "freeze in" the chemical equilibrium between these two species, i.e. we keep N_L^0 and N_H^0 fixed. We add a small amount of solute s, dN_s, and examine the change in the value of the difference, $\mu_L - \mu_H$.

Clearly, at equilibrium $\mu_L - \mu_H = 0$. However, after adding dN_s molecules of the solute, the difference $\mu_L - \mu_H$ might either be positive or negative. Equation (2.2) tells us that if $\mu_L - \mu_H$ is positive, i.e. the derivative $(\partial \Delta\mu/\partial N_s)_*$ is positive, then after restoration of the equilibrium between L and H, water molecules will "flow" from the species having the higher chemical potential to that having the lower chemical potential. In this case from L to H, or equivalently the derivative on the LHS of Eq. (2.2) must be negative (i.e. N_L decreases with the addition of the solute). If, on the other hand, $\mu_L - \mu_H$ turns out to be negative, then the derivative on the LHS of Eq. (2.2) must be positive.

In liquid water, we can identify the LLD species L with the more *structured* component, i.e. those molecules which are engaged in up to four HBs to their environment; we can say that an increase in the "structure" is equivalent to a negative value of the derivative of $\Delta\mu$ on the RHS of Eq. (2.2). Whenever the derivative on the LHS of Eq. (2.2) is positive, we will say that L is *stabilized* by the addition of s.

There are many experimental results which indicate that inert solutes do cause an increase in the structure of water. They are reviewed in Ben-Naim (1974, 2009). On the other hand, a theoretical explanation of this effect is not easy.

To the best of the author's knowledge, the only theoretical argument available on the negative sign of $(\frac{\partial\Delta\mu}{\partial N_s})_*$ — or, equivalently, on the positive sign of the derivative $(\frac{\partial N_L}{\partial N_s})_{\text{eq}}$ — is in terms of the Kirkwood–Buff theory of solutions [Ben-Naim (2006, 2009)]. This theory provides expressions for all the derivatives on the RHS of Eq. (2.2) in terms of the so-called Kirkwood–Buff integrals (KBIs). For very dilute solutions of s in water w, the result is

$$\lim_{\rho_S \to 0} \left(\frac{\partial N_L}{\partial N_S}\right)_{\text{eq}} = \rho_W x_L x_H [(G_{WH} - G_{WL})$$

$$+(G_{LS} - G_{HS})], \qquad (2.3)$$

where $\rho_W = \rho_L + \rho_H$, and x_L and x_H are the mole fractions of the two species L and H, respectively. G_{ij} are the KBIs between the species i and j.

The KBIs for the two species i and j, G_{ij}, are defined by

$$G_{ij} = \int_0^\infty [g_{ij}(R) - 1]4\pi R^2 dR, \qquad (2.4)$$

where $g_{ij}(R)$ is the (angle-averaged) pair correlation function for the species i and j, and R is the distance between the two molecules. It should be noted that the pair correlation function must be taken in the open system for which the integrand $[g_{ij}(R) - 1]$ becomes zero at large distances R. [See Ben-Naim (2006).]

In Eq. (2.3), G_{LS} and G_{HS} are KBIs between L and s, and H and s, respectively. On the other hand, G_{WH} is the KBI between *any* water molecule and an H molecule. Likewise, G_{WL} is the KBI between *any* water molecule and an L molecule. [For more details, the reader is referred to Ben-Naim (2009).]

The relation (2.3) is very general. First, it applies to any two-component system at chemical equilibrium, and to any classification procedure we have chosen for the two quasi-components, L and H. Second, because of the application of the Kirkwood–Buff theory of solutions, we do not have to restrict ourselves to any assumption of additivity for the total potential energy of the system. Furthermore, the KBIs $G_{\alpha\beta}$ depend only on the *spatial* pair correlation functions $g_{\alpha\beta}(R)$, even though we may be dealing with nonspherical particles.

Let us consider some general implications of (2.3) with regard to the conditions for the stabilization of L by s.

(1) When either x_L or x_H is very small, the whole RHS of (2.3) is small and we cannot get a large stabilization effect. If we choose, for instance, L to be strictly ice molecules, then it is likely that x_L will be small; hence, a small stabilization effect is expected. This comment is very important when one is arguing about the formation

of any *new structure*, or a structure that exists in a very low concentration in liquid water. The enhancement of any "structure" by the solute depends on the mole fraction of that component in the pure liquid which we identify as being the more structured.

(2) If we choose two components L and H which are *similar* in the sense that

$$G_{LS} \approx G_{HS}, \quad G_{WH} \approx G_{WL}, \qquad (2.5)$$

then we again end up with a small stabilization effect. Thus, in order to get the large term in Eq. (2.3), we must choose two components that are *very different*.

(3) Suppose that the product $x_L x_H$ is not too small and that L and H differ appreciably. The RHS of (2.3) will tend to zero as $\rho_W \to 0$.

All the considerations made thus far apply to any fluid. We have seen that a large stabilization effect is attainable only under very restricted conditions. Water, as one of its unique features, may conform to all the necessary conditions, leading to a relatively large stabilization effect.

Before applying what we have learned so far for the interpretation of the solvation entropy and enthalpy, we note that on the RHS of (2.3), there are two terms in the squared brackets. One, $G_{WH} - G_{WL}$, depends only on the solvent, and on the choice we have made regarding the two components. The other, $G_{LS} - G_{HS}$, depends on the relative affinity of the solute s to the two components.

For a specific choice of two components for water, L and H, we have [Ben-Naim (2006, 2009)]

$$\bar{V}_L - \bar{V}_H > 0 \quad \text{or, equivalently,} \quad G_{WH} - G_{WL} > 0. \quad (2.6)$$

The quantity $\rho_W(G_{WH} - G_{WL})$ measures the excess of *water* molecules around H as compared to the excess of water molecules around L. This means that if we can define two components which are *very different* and for which $x_L x_H$ is close to its maximum value, then we have already guaranteed one positive term for the stabilization effect which is independent of the solute.

The next question concerns the sign of $G_{LS} - G_{HS}$ for simple solutes. From the definition of the KBI, we have

$$\rho_S \Delta_{LH}^S = \rho_S(G_{LS} - G_{HS}) = \rho_S \int_0^\infty [g_{LS}(R) - 1] 4\pi R^2 dR$$

$$-\rho_S \int_0^\infty [g_{HS}(R) - 1] 4\pi R^2 dR$$

$$= \rho_S \int_0^\infty [g_{LS}(R) - g_{HS}(R)] 4\pi R^2 dR. \quad (2.7)$$

Clearly, $\rho_S \Delta_{LH}^S$, defined in Eq. (2.7), measures the overall average excess of s molecules in the neighborhood of L relative to H. From the very definition of the two components, we expect that an L molecule, characterized by an LLD, will allow more solute molecules to enter its surroundings. This is shown schematically in Figure 2.10. Therefore, Δ_{LH}^S is likely to be positive.

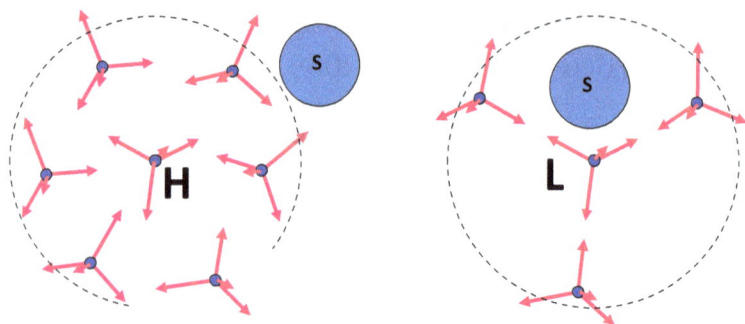

Fig. 2.10. A solute s approaching an H and an L water molecule.

Thus far, we have shown that if we choose two components in such a way that one has a relatively LLD (L), then it is likely that this component will be "stabilized" (i.e. its concentration increases) by the addition of the solute s. This argument applies for *any* fluid. It is expected, though, to be particularly large for water which has an outstanding mode of packing of water molecules. More important, however, is the peculiar and unique coupling between LLD and strong binding energy (to which we refer as the principle of water). The latter is essential to the interpretation of the large negative enthalpy and entropy of solution.

Thus, both terms on the RHS of Eq. (2.3) should be positive for any fluid for which we have classified the solvent molecules as LLD (L) and HLD species (H). Both terms are positive by definition of the LLD component. The first is related to the partial molar volumes of the two species. The second is due to the fact that the LLD component has more room in its vicinity to accommodate the solutes (Figure 2.10).

In the next subsection, we will see how this positive quantity, combined with the specific *principle* of water, provides a plausible interpretation of the entropy and enthalpy of solvation of nonpolar solutes in water. It should be noted that this effect is also realized for a hard sphere solute or, equivalently, for the creation of a cavity in water. A cavity in any liquid is a region from which solvent molecules are excluded. This excluded volume is nothing but the solvation Gibbs energy of a hard sphere in water. It was shown long ago by Pierotti (1963, 1965, 1967) and by Ben-Naim and Friedman (1967) that the solvation entropy of a hard sphere solute in water is indeed large and negative. (See also Appendix E.)

2.4.2 *The General Statistical-Mechanical Expressions for the Thermodynamics of Solvation of Nonpolar Solutes in Water*

In this subsection, we start with the general statistical-mechanical expressions for the solvation quantities of a solute *s* in water. We then interpret these quantities in terms of the *structural changes* induced in the solvent and combine the result of the previous subsection with the *principle* of water to arrive at an interpretation of the large negative entropy and enthalpy of solvation.

In the following, we will use the T, V, N ensemble, which is somewhat simpler. All the conclusions of this subsection are valid for the T, P, N ensemble as well. Starting from the definition of the chemical potential of the solute *s* in pure water, we get the expression (for a simple structureless solute in dilute solutions in water)

$$\mu_s = -k_B T \ln \frac{Q(T, V, N; 1s)}{Q(T, V, N)}$$

$$= k_B T \ln \rho_s \Lambda_s^3 - k_B T \ln \langle \exp[-\beta B_s] \rangle_0, \qquad (2.8)$$

where k_B is the Boltzmann constant, T the absolute temperature, and $Q(T, V, N)$ the canonical partition function of a system of N solvent molecules at volume V and temperature T; $Q(T, V, N; 1s)$ is the partition function for the same solvent at T, V, N but with an added solute s, ρ_s the number density of the solute, Λ_s^3 the momentum partition function of s, and B_s the "binding energy" of s to all solvent molecules. B_s is defined by

$$B_s = \sum_{i=1}^{N} U(R_s, X_i), \qquad (2.9)$$

with R_s the locational vector of s, and X_i describe the orientation and location of the ith solvent molecule. Note that in Eq. (2.9), we define the binding energy of the solute for a system with pairwise-additive interactions. For more general systems, see Ben-Naim (1974, 2006). The average $\langle \, \rangle_0$ is an average in the T, V, N ensemble with the distribution of configurations of the *pure* solvent

$$P(X^N) = \frac{\exp[-\beta U_N(X^N)]}{\int \cdots \int dX^N \exp[-\beta U_N(X^N)]}, \qquad (2.10)$$

where $\beta = (k_B T)^{-1}$.

Thus, the Helmholtz energy of solvation is given by

$$\Delta A_s^* = -k_B T \ln(\exp[-\beta B_s])_0. \qquad (2.11)$$

The Gibbs energy of solvation has the same form with reinterpretation of the average quantity in Eq. (2.11) in the T, P, N ensemble. Equation (2.8) may be written as

$$\mu_s = \mu_s^* + k_B T \ln \rho_s \Lambda_s^3. \qquad (2.12)$$

This equation has a simple interpretation in terms of two steps of adding the solute s to the liquid (l). First, we add the solute s to a fixed point in the liquid. The corresponding change in the Helmholtz energy is μ_s^*, referred to as the pseudochemical potential. This quantity involves mainly the work of *coupling* the solute to the solvent. [For details, see Ben-Naim (2006).] The second term is referred to as the liberation Helmholtz energy. It is the change in the Helmholtz energy when the solute is released from its fixed position in the liquid. Note that the quantity $\rho_s \Lambda_s^3$ must be very small for the validity of the classical limit of statistical mechanics, and hence $k_B T \ln \rho_s \Lambda_s^3$ is always a negative quantity. [See Hill (1956).]

We note here that the second term on the RHS of Eq. (2.12) consists of three contributions to the chemical potential. One is due to the accessibility of the entire volume V. The second is due to the attainability of the Maxwell–Boltzmann distribution of momenta. The third is due to assimilation of the added solute into the N_s solute molecules [see Ben-Naim (2006)].

Taking the temperature derivative of ΔA_s^* and using the identity $\Delta A_s^* = \Delta E_s^* - T\Delta S_s^*$, we get the final expressions [for details, see Appendix F and Ben-Naim (1992, 2006, 2009)]

$$\Delta S_s^* = k_B \ln \langle \exp[-\beta B_s] \rangle_0$$

$$+ \frac{1}{T} \left[\langle B_s \rangle_s + (\langle U_N \rangle_s - \langle U_N \rangle_0) \right], \quad (2.13)$$

$$\Delta E_s^* = \langle B_s \rangle_s + (\langle U_N \rangle_s - \langle U_N \rangle_0). \quad (2.14)$$

Note carefully that the symbol $\langle \rangle_0$ stands for an average (here in the T, V, N ensemble) with the distribution of the configurations of the *pure* solvent molecules [Eq. (2.10)], i.e. *before* the insertion of the solute. On the other hand, the symbol $\langle \rangle_s$ stands for an average with a *conditional* distribution of the configurations of the solvent molecules given a solute s at some fixed point in the solvent, i.e.

$$P(X^N | R_s) = \frac{\exp\left(-\beta U_N - \beta B_s\right)}{\int dX^N \exp\left(-\beta U_N - \beta B_s\right)}. \quad (2.15)$$

Note that the term $\langle U_N \rangle_s - \langle U_N \rangle_0$ is common to ΔS_s^* and ΔE_s^*. (In the T, P, N ensemble, ΔE_s^* is replaced by the enthalpy of solvation ΔH_s^*.) This term is simply the change in the total interaction energy among all solvent molecules induced by placing a solute s at some fixed position in the liquid. It may also be reinterpreted as structural changes induced in the solvent, or reorganization of the solvent or redistribution of water molecules. The general case is presented in Appendix G, using the binding energy distribution function. Here, we note that while this term features in both ΔS_s^* and ΔE_s^* (as well as in ΔH_s^*) it does not feature in ΔA_s^* (as well as in the Gibbs energy of solvation,

ΔG_s^*). This is an important observation, the implication of which will be discussed in the next subsection.

Here, we note that if we accept the interpretation of the term $\langle U_N \rangle_s - \langle U_N \rangle_0$ as structural changes in the solvent induced by the solute s, we can conclude from Eqs. (2.10), (2.12) and (2.13) the following: whatever the value of the quantity $\langle U_N \rangle_s - \langle U_N \rangle_0$ is, this quantity can affect both the entropy and the enthalpy of solvation of the solute s. It cannot affect the Gibbs energy of solvation of the solute. In the next subsection, we discuss a simpler version of this conclusion using the mixture model approach to liquid water.

2.4.3 Application of the Two-Structure Model for Water; the Low- and High-Local-Density Components

In Appendix F, the average in Eq. (2.11) is rewritten in terms of the distribution function for the binding energy. Similarly, by taking the temperature derivative of ΔA_s^*, we can express both ΔS_s^* and ΔE_s^* in terms of the distribution of the binding energies for solvent molecules. Here, we present a simple argument based on the mixture model approach to liquid water. This model was originally considered an *ad hoc* model for water, but later it was shown that this model could be made exact [Ben-Naim (1992, 2009)]. For our purpose, we use a classification of all water molecules into two *species*: the low-local-density (L) and high-local-density (H). We also identify the L species as the more *structured* one, and hence the larger the concentration of L, the larger the structure of the water.

For any given temperature, pressure and N_s solute molecules, we can view the Gibbs energy of the system as a function of either the four variables (T, P, N_w, N_s) or the five variables (T, P, N_L, N_H, N_s) where N_L and N_H are connected by the equilibrium condition

$$\mu_L - \mu_H. \tag{2.16}$$

We now write the chemical potential and the partial molar entropy and enthalpy of the solute as

$$\mu_s = \left(\frac{\partial G}{\partial N_s}\right)_{N_w} = \left(\frac{\partial G}{\partial N_s}\right)_{N_L, N_H}$$
$$+ (\mu_L - \mu_H)\left(\frac{\partial N_L}{\partial N_s}\right)_{N_w}, \tag{2.17}$$

$$\bar{S}_s = \left(\frac{\partial S}{\partial N_s}\right)_{N_w} = \left(\frac{\partial S}{\partial N_s}\right)_{N_L, N_H}$$
$$+ (\bar{S}_L - \bar{S}_H)\left(\frac{\partial N_L}{\partial N_s}\right)_{N_w}, \tag{2.18}$$

$$\bar{H}_s = \left(\frac{\partial H}{\partial N_s}\right)_{N_w} = \left(\frac{\partial S}{\partial N_s}\right)_{N_L, N_H}$$
$$+ (\bar{H}_L - \bar{H}_H)\left(\frac{\partial N_L}{\partial N_s}\right)_{N_w}. \tag{2.19}$$

In each of these equations, we first use the definition of the partial molar quantity, employing the independent variables (T, P, N_w, N_s). T and P are kept constant and are omitted from these equations. In the second step, we use the independent variables T, P, N_L, N_H, N_s. We first write the derivative at constant values of N_L and N_H, and then we let the system *relax* to a new equilibrium state with respect to the reaction $L \rightleftharpoons H$.

Because of the equilibrium condition (2.16), the expression for the chemical potential in Eq. (2.17) reduces to

$$\mu_s = \left(\frac{\partial G}{\partial N_s}\right)_{N_w} = \left(\frac{\partial G}{\partial N_s}\right)_{N_L, N_H}. \qquad (2.20)$$

This means that the work associated with introducing one s molecule (or one mole of molecules) into the system is the same whether or not we *freeze* the conversion between L and H. This statement is valid for an infinitesimal change in the concentration of the solute. For the effect of a finite addition of solute molecules, see Appendix H.

On the other hand, both $\bar{S}_s$ and $\bar{H}_s$ will in general have two contributions. One is due to introducing the solute into the system while holding N_L and N_H fixed. The other is due to the change in the average number of L and H induced by the addition of the solute s. We can refer to this term as the effect of *structural* changes induced by the solute on the solvent. Figure 2.9 shows the two steps of adding a solute to the system.

Had we worked in the T, V, N ensemble, i.e. addings solute at constant volume instead of constant pressure, Eqs. (2.18)–(2.20) would be interpreted as being at constant T and V rather than T and P. In addition, the partial molar energy of the solute would be

$$\bar{E}_s = \left(\frac{\partial E}{\partial N_s}\right)_{N_w} = \left(\frac{\partial E}{\partial N_s}\right)_{N_L, N_H} + (\bar{E}_L - \bar{E}_H)\left(\frac{\partial N_L}{\partial N_s}\right)_{N_w}.$$
$$(2.21)$$

Here, all the derivatives are at constant T and V.

So far, we have discussed the partial molar quantities of the solute s. Clearly, the same reasoning can be applied to the pseudochemical potential and to the other pseudopartial molar quantities.

For the solvation process as defined in Appendix A, we get the results

$$\Delta\mu_s^* = \Delta\mu_s^{*(\mathrm{fr})}, \tag{2.22}$$

$$\Delta S_s^* = \Delta S_s^{*(\mathrm{fr})} + (\bar{S}_L - \bar{S}_H)\left(\frac{\partial N_L}{\partial N_s}\right)_{N_w}, \tag{2.23}$$

$$\Delta H_s^* = \Delta H_s^{*(\mathrm{fr})} + (\bar{H}_L - \bar{H}_H)\left(\frac{\partial N_L}{\partial N_s}\right)_{N_w}, \tag{2.24}$$

and similarly for the energy of solvation in the T, V, N ensemble

$$\Delta E_s^* = \Delta E_s^{*(\mathrm{fr})} + (\bar{E}_L - \bar{E}_H)\left(\frac{\partial N_L}{\partial N_s}\right)_{N_w}. \tag{2.25}$$

The first terms on the RHS of each of these equations are the change of the thermodynamic quantity for introducing s into a frozen in (fr) system, i.e. when the conversion between L and H is forbidden.

In Subsection 2.4.1, we showed that the derivative $(\partial N_L/\partial N_s)_{\mathrm{eq}}$ is positive for any classification of water molecules into LLD, and HLD. For a general liquid, one would expect that the difference in the partial molar energies $\bar{E}_L - \bar{E}_H$ is *positive*. This is because LLD is, in general, correlated with weaker local interaction energy. Therefore, in general, we expect that the sign of this term will be *positive*.

However, for water we have the characteristic correlation between LLD and *stronger binding energy*, and hence we expect that $\bar{E}_L - \bar{E}_H$ will be negative (the same is true of the enthalpy difference $\bar{H}_L - \bar{H}_H$. Therefore, we expect a negative contribution due to this term to the solvation energy (as well as the enthalpy) of the nonpolar solute.

The same quantity appears in the entropy of solvation. Within this simple classification into L and H species, we can write at equilibrium

$$\Delta\mu = \mu_L - \mu_H = 0 \qquad (2.26)$$

or, equivalently,

$$\mu_L - \mu_H = (\bar{E}_L - T\bar{S}_L) - (\bar{E}_H - T\bar{S}_H) = 0. \qquad (2.27)$$

This equality may be referred to as the exact *entropy–energy* compensation.

Hence,

$$(\bar{E}_L - \bar{E}_H)\left(\frac{\partial N_L}{\partial N_s}\right)_{eq} = T(\bar{S}_L - \bar{S}_H)\left(\frac{\partial N_L}{\partial N_s}\right)_{eq}. \qquad (2.28)$$

Thus, having an explanation for the *enhancement* of the structure of water by a solute s also provides an explanation for the large and negative entropy, and energy (as well as enthalpy) of solvation. It was estimated long ago that this iceberg formation contributes about half of the entropy of solvation of methane in water. (See Subsection 2.4.5.) We conclude that structural enhancement (we sometimes use the concept of iceberg formation to mean any enhancement of the structure of the solvent, specifically, we believe that the enhancement is only of structures that already exist in liquid water) induced by a nonpolar solute can explain the large negative entropy and enthalpy of solvation provided that neither x_L nor x_H is too small (or too close to unity). Note that the main argument used to explain the large negative entropy of solvation is essentially the same as the one used in the explanation of the anomalous temperature dependence of the volume of water [see Ben-Naim (2009)]. The next question is whether this iceberg formation also explains the Helmholtz or Gibbs energy of solvation, i.e. the solvation of hydrophobic solutes.

It is very clear from Eqs. (2.22)–(2.24) that whatever the structural changes induced in the solvent are, the entropy and enthalpy of the solvation will be affected. We do not know how much such structural changes affect the entropy and enthalpy of solvation, but we know at least the sign of these contributions. However, because of the exact entropy–enthalpy (or energy) compensation [Eq. (2.28)], the Gibbs energy of solvation will not be affected. This conclusion was reached here within the simple mixture model approach. However, the validity of this conclusion holds even in the most general case. It follows from Eqs. (2.11), (2.13) and (2.14), when the term $\langle U_N \rangle_s - \langle U_N \rangle_0$ may be reinterpreted in terms of structural changes induced in the solvent, whatever these structural changes may be.

2.4.4 An Exact Argument Using a Hypothetical Hard Point Particle

It is well known that the solvation Gibbs energy of a hard sphere is equal to the cavity work in a liquid; see Ben-Naim

(2006). A cavity of radius $\sigma/2$ can be formed by a hard solute of diameter zero. We will refer to such a particle as a hard point. The solvation Gibbs energy of such a hard point is thus

$$\Delta G_\bullet^* = -k_B T \ln (1 - \rho V_\bullet^{EX}), \qquad (2.29)$$

where $V_\bullet^{EX}$ is the excluded volume of the hard point. Since $\rho V_\bullet^{EX}$ is the probability of finding the volume $V_\bullet^{EX}$ occupied and $1 - \rho V_\bullet^{EX}$ is the probability of finding $V_\bullet^{EX}$ empty, it follows that $\Delta G_\bullet^*$ is always positive. Clearly, this relation can be applied to any liquid. The volume $V_\bullet^{EX}$ is roughly the volume of the region impenetrable by the solvent molecules. It will be proportional to the actual volume of the solvent molecules. Note that the work required to create a cavity of size zero at some fixed position is always zero; here, a hard point always creates a cavity of finite size provided that the solvent molecules can be assigned a finite hard core diameter.

Equation (2.29) suggests that we use the hard point particle as our test solute to compare the solvation thermodynamics of this solute in different solvents. We immediately see from the equation that our test solute will be more soluble in a liquid for which the quantity $\rho V_\bullet^{EX}$ is smaller. In other words, decreasing either the density or the size of the solvent particles causes a decrease in $\Delta G_\bullet^*$ or an increase in solubility. Here, we refer to solubility from an ideal gas phase. For real solutes, the attractive part of the solute–solvent interaction contributes significantly to ΔG_S^*.

The corresponding solvation entropy and enthalpy of the hard point are

$$\Delta S_S^* = - \left(\frac{\partial \Delta G_{\bullet}^*}{\partial T} \right)_P = k_B \ln (1 - \rho V_{\bullet}^{EX})$$

$$- \frac{k_B T V_{\bullet}^{EX}}{1 - \rho V_{\bullet}^{EX}} \left(\frac{\partial \rho}{\partial T} \right)_P \qquad (2.30)$$

$$\Delta H_{\bullet}^* = \Delta G_{\bullet}^* + T \Delta S_{\bullet}^* = \frac{-k_B T^2 V_{\bullet}^{EX}}{1 - \rho V_{\bullet}^{EX}} \left(\frac{\partial \rho}{\partial T} \right)_P. \qquad (2.31)$$

Note that since the hard point has no "soft" interaction, the average binding energy $\langle B_S \rangle_S$ is zero (as for any hard particle of any size). Therefore, all of $\Delta H_{\bullet}^*$ must be due to "structural changes" induced in the solvent. This can be interpreted in terms of a relaxation term using any method for classifying the solvent molecules into quasicomponents. The following important conclusion can be derived from Eqs. (2.30) and (2.31): for normal liquids, the density always decreases with temperature, i.e. $\partial \rho / \partial T < 0$, and hence $\Delta H_{\bullet}^*$ is positive. In water, we know that there exists a region between $0°C$ and $4°C$ for which $\partial \rho / \partial T > 0$. Therefore, $\Delta H_{\bullet}^*$ is negative. Since this must be due to structural changes in water, we conclude that the unique temperature dependence of the density is intimately related to the negative relaxation part of $\Delta H_{\bullet}^*$. Because of the exact compensation effect, the same conclusion applies to the relaxation part of $\Delta S_{\bullet}^*$.

This exact relationship has been derived here for a hypothetical hard point particle. For a real solute, essentially the same conclusion was inferred from the discussion based on the Kirkwood–Buff theory. Note also that the relaxation term in $\Delta H_{\bullet}^{*}$ appears in $T \Delta S_{\bullet}^{*}$ too. Therefore, in the formation of the combination $\Delta H_{\bullet}^{*} - T \Delta S_{\bullet}^{*}$, this part cancels out.

2.4.5 An Approximate Estimate of the Contribution of Structural Changes to the Solvation Entropy

We have seen that the solvation Gibbs energy of an inert solute depends only on the $\Delta H_{s}^{\mathrm{fr}}$ and $\Delta S_{s}^{\mathrm{fr}}$, and not on the relaxation terms. Hence, we write

$$\Delta \mu_{s}^{*} = \Delta H_{s}^{\mathrm{fr}} - T \Delta S_{s}^{\mathrm{fr}}. \qquad (2.32)$$

It is known that the solubility of argon in ethylene glycol is as low as in water (see Figure 2.11). For instance, at $t = 20°C$, the solvation Gibbs energy of argon in water (w) and in ethylene glycol (EG) are nearly the same, i.e.

$$\Delta \mu_{s}^{*}(\text{in } w) = \Delta \mu_{s}^{*}(\text{in EG}). \qquad (2.33)$$

We now write the solvation entropy in the two solvents,

$$\Delta S_{s}^{*}(\text{in } w) = \Delta S_{s}^{\mathrm{fr}}(\text{in } w) + \Delta S_{s}^{r}(\text{in } w), \qquad (2.34)$$

$$\Delta S_{s}^{*}(\text{in EG}) = \Delta S_{s}^{\mathrm{fr}}(\text{in EG}) + \Delta S_{s}^{r}(\text{in EG}), \qquad (2.35)$$

and similarly for the enthalpy of solvation,

$$\Delta H_{s}^{*}(\text{in } w) = \Delta H_{s}^{\mathrm{fr}}(\text{in } w) + \Delta H_{s}^{r}(\text{in } w), \qquad (2.36)$$

$$\Delta H_{s}^{*}(\text{in EG}) = \Delta H_{s}^{\mathrm{fr}}(\text{in EG}) + \Delta H_{s}^{r}(\text{in EG}). \qquad (2.37)$$

We now assume that $\Delta H_{s}^{r}(\text{EG}) \approx 0$. There are two reasons for this assumption. First, we believe that the structural change of the type that occurs in water is peculiar to water

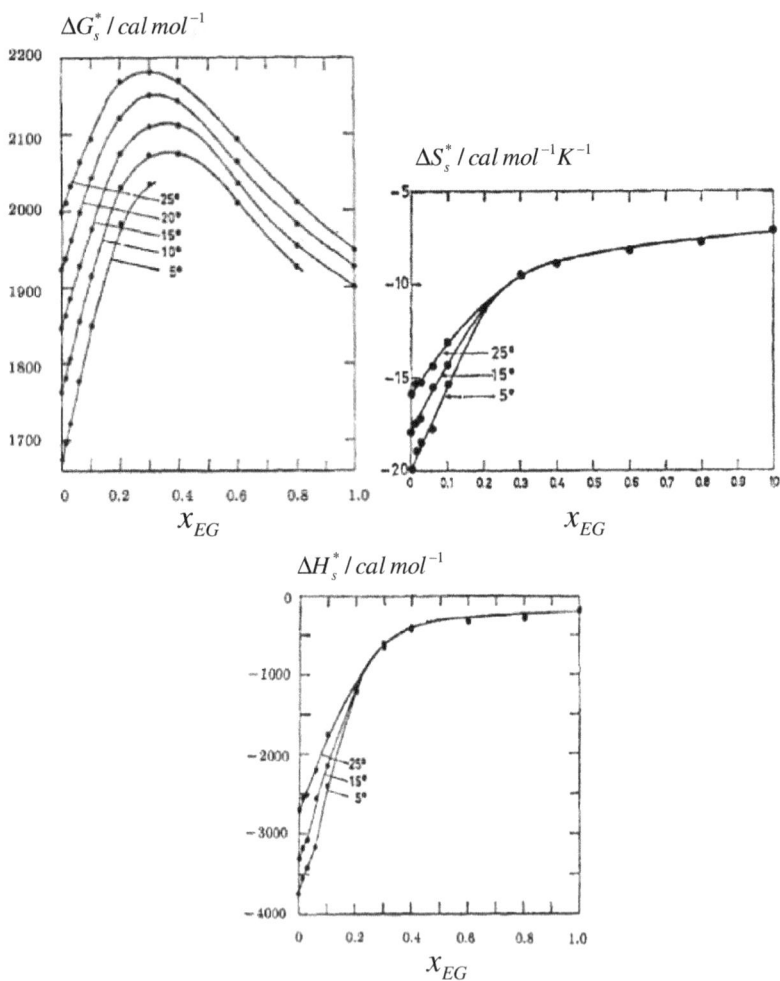

Fig. 2.11. Solvation thermodynamics of argon in mixtures of water and ethylene glycol at three temperatures.

and probably arises from the existence of the open structure of water. Secondly, if there were any appreciable structural changes of this kind in ethylene glycol, these should have shown up in the total enthalpy and the entropy of solvation. The experimental data show that ΔH_s^*(EG) has a "normal" value, i.e. of the same order of magnitude as in other organic solvents. We therefore put $\Delta H_s^r(\text{EG}) \approx 0$ and write the approximations

$$\Delta H_s^*(\text{in EG}) = \Delta H_s^{\text{fr}}(\text{in EG}), \qquad (2.38)$$

$$\Delta S_s^*(\text{in EG}) = \Delta S_s^{\text{fr}}(\text{in EG}). \qquad (2.39)$$

Since ΔH_s^{fr} arises mainly from the total interaction energy of argon with the solvent, we can assume that $\Delta H_s^{\text{fr}}(\text{in } w)$ and $\Delta H_s^{\text{fr}}(\text{in EG})$ are of the same order of magnitude; about -400 cal/mol. The experimental value for $\Delta H_s^*(\text{in } w)$ is much more negative than this, so that we can put

$$\Delta H_s^*(\text{in } w) \approx \Delta H_s^r(\text{in } w). \qquad (2.40)$$

From the above assumptions, we arrive at the following estimation (valid for the particular temperature chosen above):

$$\Delta S_s^f(\text{in } w) \approx \Delta S_s^*(\text{in EG}) \approx -7\,\text{cal}\,\text{mol}^{-1}K^{-1}, \quad (2.41)$$

$$\Delta S_s^r(\text{in } w) \approx \Delta S_s^*(\text{in } w) - \Delta S_s^{\text{fr}}(\text{in } w)$$

$$\approx -10\,\text{cal}\,\text{mol}^{-1}K^{-1}. \qquad (2.42)$$

This is an interesting result, since we now have an idea of the order of magnitude of the two terms that contribute to the entropy of solvation. The conclusion is that about half of the anomalous large and negative entropy of solvation comes from the "frozen-in" part, and the other half from structural changes induced in the solvent. Since the only part that affects the solubility comes from the frozen-in terms and since ΔH_s^{fr} is assumed to be relatively small, we conclude that about $-7\,\mathrm{cal\,mol^{-1}}K^{-1}$ from the total entropy of solvation is responsible for the low solubility of gases in water.

2.4.6 Generalization for HϕO Interaction and for Any Reaction

In the previous subsections, we showed that when the solute s affects the distribution of "species" (or quasicomponents) in the solvent, we might expect contribution of this effect to the solvation entropy and the solvation energy (or enthalpy). We briefly show here that the same conclusion holds for the hydrophobic interaction, i.e. for the PMF between two hydrophobic molecules, and for any chemical reaction, in particular the reaction of folding–unfolding in aqueous solutions.

Consider the process of bringing two hydrophobic molecules — say, methane molecules — from infinite separation to the distance R (Figure 2.12). The PMF is simply the Gibbs energy change for this process (T, P and the number of solvent molecules are kept constant).

$$\Delta G(\infty \to R) = \Delta G(R) = G(R) - G(\infty). \quad (2.43)$$

Here, we use the notation $\Delta G(R)$ for the process $\infty \to R$. One can always write the PMF in two terms:

$$\Delta G(R) = \Delta U(R) + \delta G(R), \quad (2.44)$$

Fig. 2.12. The process of bringing two solute particles from infinity to the distance R.

where $\Delta U(R)$ is the change of energy for the process of bringing the two molecules to R in vacuum, and $\delta G(R)$ may be referred to as the solvent-induced contribution to $\Delta G(R)$. The hydrophobic *pair interaction* is sometimes identified with the PMF [see Ben-Naim (1974, 2009)]. However, since $\Delta U(R)$ is presumed to be independent of the solvent, the more appropriate quantity to be referred to as hydrophobic interaction is $\delta G(R)$ [Ben-Naim (1974)].

One can easily show that $\delta G(R)$ may be obtained by differences in solvation Gibbs energies:

$$\delta G(R) = \Delta G_D^*(R) - 2\Delta G_M^*, \qquad (2.45)$$

where ΔG_M^* is the solvation Gibbs energy of the monomer — say, methane — and $\Delta G_D^*(R)$ is the solvation Gibbs energy of the two monomers being at a fixed separation R. (It should be noted that the above definition of the quantity $\Delta G_D^*(R)$ is valid only when we use the definition of the solvation process as defined in Figure 2.4. It will be awkward to use the "excess chemical potential" for this quantity. (See also Appendix A.)

The relationship (2.45) can also be deduced from the cyclic reaction shown in Figure 2.13. One can carry out the process $\infty \rightarrow R$ in vacuum and in the solvent. Subtract the two Gibbs energies and we find that

$$\Delta G^l(R) - \Delta G^{ig}(R) = \Delta G(R) - U(R) = \delta G,$$
$$= \Delta G_D^*(R) - 2\Delta G_M^*. \qquad (2.46)$$

Thus, we see that the $H\phi O$ pair interaction [in either the sense of the PMF, $\Delta G(R)$, or the sense of $\delta G(R)$] is related to differences in solvation Gibbs energies. The entropy and

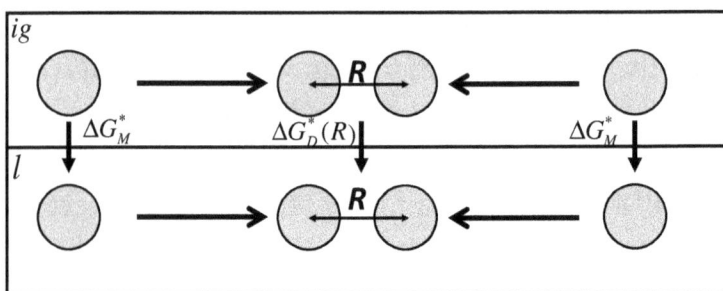

Fig. 2.13. Relation between solvation and hydrophobic interaction.

the enthalpy of the process $\infty \to R$ are

$$\Delta S(R) = -\frac{\partial \Delta G(R)}{\partial T} = \Delta S_D^*(R) - 2\Delta S_M^*, \quad (2.47)$$

$$\Delta H(R) = \Delta U(R) + \Delta H_D^*(R) - 2\Delta H_M^*. \quad (2.48)$$

Thus, the conclusion we have reached in the previous subsections regarding the contribution of structural changes induced in the solvent to ΔG_S^* (i.e. for the solvation process) also holds for the process $\infty \to R$.

We can generalize this result to any process in a liquid. For instance, the folding of a protein may be written as

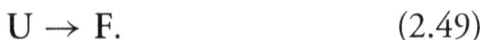

$$U \to F. \quad (2.49)$$

The standard Gibbs energies of this reaction can be written as

$$\Delta G^{0l}(U \to F) = \Delta G^{0ig}(U \to F) + \Delta G_F^* - \Delta G_U^*. \quad (2.50)$$

This relationship may be deduced from the cyclic process in Figure 2.14. Again, the standard Gibbs energy in vacuum or in an ideal gas phase is independent of the solvent. Hence,

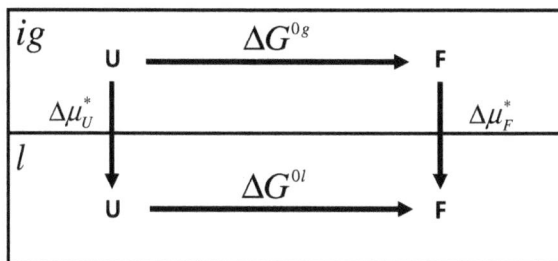

Fig. 2.14. Relation between solvation and standard GE of the reactions.

ΔG^{0l} is related to the difference in the solvation Gibbs energies of the U and the F forms.

When a protein folds or unfolds, there might be some changes in the structure of the solvent. These changes are caused by the solvation properties of all the groups which are exposed to the solvent. Some groups might cause a positive enhancement of the structure, and some negative. For this reason it is premature to claim that protein folding is associated with enhancement or destruction of the structure of water. However, whatever the changes in the structure might be, they will contribute to the standard entropy and the enthalpy of the reaction, but they cannot contribute to the standard Gibbs energy of the reaction.

2.5 Does the Enhancement of the Structure of Water Explain the Large Positive Partial Molar Heat Capacity of Nonpolar Solutes in Water?

It is often stated that iceberg formation also explains the partial molar heat capacity of a nonpolar solute in water.

The argument, originally given by Frank and Evans, is as follows:

(i) The high value of the heat capacity of water is due to the melting of the ice-like component in pure water as we heat the mixture.

(ii) It is known that a solute s increases the structure of water. This means that the concentration of the ice-like component *increases* with the addition of the solute s.

Assuming that (i) and (ii) are true, it follows that the large positive value of the partial molar heat capacity of s in water can be explained by the fact that "more of the ice-like component is available for melting."

We present here a simple model showing that this conclusion is not valid. The partial molar heat capacity of a solute does not necessarily increase with the concentration of the "ice-like" component. This topic is discussed in detail in Ben-Naim (2009).

Here, we present a simple argument based on a simple mixture model approach to liquid water. Furthermore, the argument here is based on the assumption that the two components form a symmetrical ideal solution. It was shown that this assumption is in general not valid for the mixture model when applied to explain the properties of water [see Ben-Naim (2006)].

Suppose that the two components H and L form a symmetrical ideal solution, in which case the chemical potentials of the two components may be written as

$$\mu_H = \mu_H^p + RT\ln x_H, \quad (2.51)$$

$$\mu_L = \mu_L^p + RT\ln x_L, \quad (2.52)$$

where μ_H^p and μ_L^p are the chemical potentials of pure H and L, respectively, and x_H and x_L are the corresponding mole fractions, defined by

$$x_H = \frac{N_H}{N}, \quad x_L = 1 - x_H = \frac{N_L}{N}, \qquad (2.53)$$

with N_H and N_L the average number of H and L molecules at equilibrium and $N = N_L + N_H$.

At equilibrium, we have

$$\mu_L = \mu_H \qquad (2.54)$$

or, equivalently,

$$0 = \mu_L - \mu_H = \mu_L^p - \mu_H^p = RT \ln \frac{x_H}{x_L}. \qquad (2.55)$$

Taking the temperature derivative of the last equation, we get

$$\frac{\partial x_L}{\partial T} = \frac{(H_L^p - H_H^p) x_L x_H}{RT^2}. \qquad (2.56)$$

For water, we can choose the component L to be the more hydrogen-bonded species, and hence $H_L^p - H_H^p < 0$. Therefore, the mole fraction of L will *decrease* upon raising the temperature.

The heat capacity of water can be obtained by taking the temperature derivative of the enthalpy of the system. For ideal solutions, we have

$$H = N_L H_L^p + N_H H_H^p \qquad (2.57)$$

and

$$C_P = N_L C_{PL}^p + N_H C_{PH}^p + (H_L^p - H_H^p) \left(\frac{\partial N_L}{\partial T} \right)_{eq},$$

(2.58)

where C_{PL}^p and C_{PH}^p are the partial molar heat capacities of the pure components L and H. The first two terms on the RHS of Eq. (2.58) are expected to be relatively small. The third term on the RHS will be referred to as the contribution to the heat capacity of water due to the "melting" of the L component. This term is referred to as the *relaxation* term, and is responsible for the large heat capacity of water:

$$C_P^{\text{relax}} = (H_L^p - H_H^p) \left(\frac{\partial N_L}{\partial T} \right)_{eq}.$$

(2.59)

Combining Eqs. (2.56) and (2.60), we get

$$C_P^{\text{relax}} = N \frac{\left(H_L^p - H_H^p \right)^2 x_L x_H}{RT^2}.$$

(2.60)

Clearly, this is always a positive quantity and it is larger the larger $H_L^p - H_H^p$ is and when x_L and x_H are close to ½. This is a reasonably satisfactory explanation for the large heat capacity of liquid water. It can already be seen from this equation that C_P^{relax} is, in general, not larger the larger the concentration of the ice-like form is.

The partial molar heat capacity of the solute s is obtained by taking the derivative of C_P with respect to N_s. However, we expect that the main contribution to the partial molar heat capacity of s will come from the derivative of the relaxation term (i.e. the third term on the RHS). Thus, we have approximately

$$C_{P,S} \approx \frac{\partial C_P^{relax}}{\partial N_s} = \frac{N(H_L^p - H_H^p)^2}{RT^2} \frac{\partial}{\partial N_s}(x_L x_H)$$

$$= \frac{N(H_L^p - H_H^p)^2}{RT^2}(x_H - x_L)\left(\frac{\partial x_L}{\partial N_s}\right).$$

$$(2.61)$$

The significance of the last equation is as follows. We assume that the "melting" of the L component is responsible for the large value of the heat capacity of water. However, even if the concentration of the L component increases, i.e. $\partial x_L/\partial N_s$ is positive, it is not clear that the "more icebergs" available necessarily means "more melting." As can be seen from Figure 2.15 and Eq. (2.61), the partial molar heat capacity of the solute depends on the difference $x_H - x_L$,

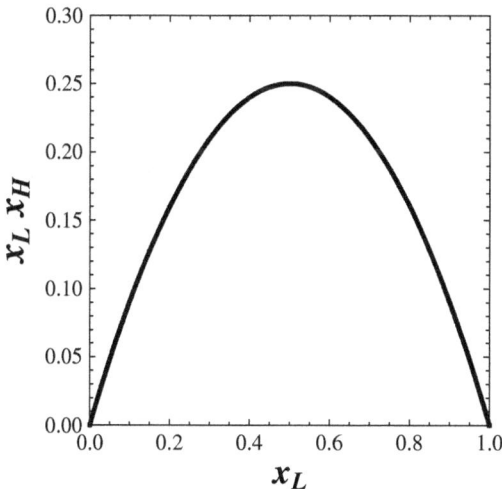

Fig. 2.15. The product $x_L x_H$ as a function of x_L.

i.e. if we have $x_H - x_L > 0$, then the whole term on the RHS of (2.61) will be positive. On the other hand, if $x_H - x_L < 0$, then this term will be negative.

Sometimes the large positive partial molar heat capacity is said to be a "signature" of the hydrophobic effect. The term "signature" usually implies a unique identifier. This is certainly true for the signature of a person on an official document. In the case of ΔC_p^*, this term might be misleading. The large positive value of ΔC_p^* is not unique to solvation of nonpolar solutes. The partial molar heat capacity of a hydrophilic solute can also be large and positive. Therefore, when observing a large positive change in heat capacity in a process such as protein folding or unfolding, one cannot conclude that the hydrophobic effect is dominant in this process.

In concluding this subsection, we can say that there is no simple relationship between ΔC_p^* and structural changes induced in a solvent. ΔC_p^* can be either positive or negative in processes inducing either enhancement or breakdown of the structure of water.

Finally, we point out that the enhancement of the structure of water by a simple solute by itself does not provide an explanation for the large positive partial molar heat capacity of the solutes in water.

2.6 Conclusion

We have seen that iceberg formation was conjectured to explain the large negative entropy of solvation. However, before using this idea to explain other phenomena, one must first show that an inert solute does enhance the structure of water. Once this is done, one can explain both the solvation entropy and the solvation enthalpy of the solute in water. The enhancement of the structure of water cannot explain the Gibbs energy of solvation, or the large partial molar heat capacity of a solute in water.

Within the simple classification into L and H species, it is clear that at equilibrium, whatever the contribution of the structural enhancement to $T\Delta S_s^*$ is, the same is true of the contribution to ΔH_s^* (or ΔE_s^*). This is a result of the exact entropy–enthalpy compensation. In other words, when one writes ΔG_s^* as $\Delta H_s^* - T\Delta S_s^*$, any structural changes affecting ΔH_s^* will be exactly compensated for with the effect on $T\Delta S_s^*$. This invalidates the conclusion based on the two correct statements (i) and (ii), in Section 2.5.

In more general terms, we refer again to Eqs. (2.13) and (2.14). It is seen that the *conditional averages* feature in both ΔH_s^* (or ΔE_s^*) and ΔS_s^*. On the other hand, ΔG_s^* (as well as ΔA_s^*) contains only an average with the distribution of configurations of the *pure* solvent [this is denoted $\langle \ \rangle_0$ in (3.4)]. Thus, whatever the structural changes induced by the solute molecule, it might affect both ΔH_s^* and ΔS_s^*, but it cannot affect ΔG_s^*. The latter depends only on the distribution of the pure solvent, i.e. before it is modified by the *condition* of introducing the solute into the system. It should be noted that the conclusion of this section is independent of the *sign* of the reorganization of the solvent molecules.

In conclusion, Frank and Evans did not explain why a simple solute will form icebergs. They certainly did not provide an explanation for the hydrophobic effects (this is true of both hydrophobic solvation and hydrophobic interaction). Therefore, all statements about the explanation of the hydrophobic effects by Frank and Evans are nothing but an unfounded myth.

3

From Schellman's Experiments to Fersht's Hydrogen Bond Inventory Argument

3.1 Abstract

"In the beginning" HBs reigned supreme in biochemical processes. The energy of an HB is about ten times larger than the typical van der Waals interaction energy. Therefore, it is almost natural to assume that the HB will play a major role in protein stability. This is certainly true of the α-helix and the β-sheets, (Figure 3.1), and of course in the formation of base pairs in DNA (Figure 3.2). Then came a devastating blow to the role of hydrogen bonding encapsulated in the so-called *hydrogen bond inventory argument*. In essence, this argument states that the contribution of the HB to biochemical processes is insignificant. Why? Because of some cancelations between the HB formed between the solutes and between the solutes and water molecules. This (erroneous) conclusion was the main motivation for Kauzmann to suggest that the hydrophobic effects rather than HBs might be the more important factor in protein stability. We will see that the

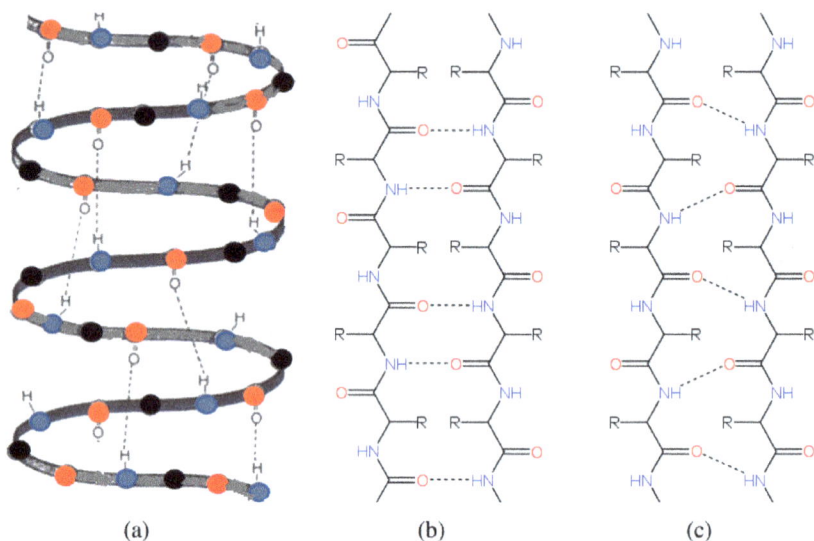

Fig. 3.1. Hydrogen bonds in (a) alpha-helix, (b) anti-parallel, and (c) parallel beta-sheet.

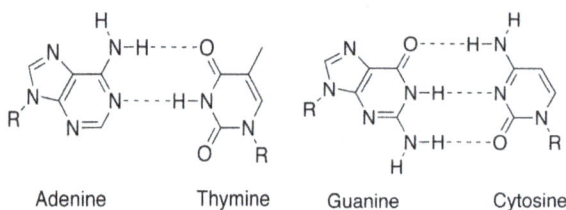

Adenine   Thymine   Guanine   Cytosine

Fig. 3.2. Hydrogen bonds in base pairs.

claim of the insignificance of HBs in biochemical processes is nothing but a myth, which originated from Schellman's experiments and evolved into Fersht's erroneous hydrogen bond inventory argument.

3.2 The Origin of the Myth and Some Historical Notes

"In the beginning," HBs were very important in biochemistry.

A convenient point to begin at is Pauling's book *The Nature of the Chemical Bond*. In the first two editions of this book, Pauling discussed the HB in water and in ice, and in many small organic compounds, both intermolecular and intramolecular HBs. In these early editions, there is no mention of proteins or nucleic acids. In the second edition, a chapter on HBs ends with some estimates of the HB energies and HB distances. The third edition contains two new sections, on HBs in proteins and HBs in nucleic acids. In these sections, Pauling discusses the role of HBs in the α-helix and the β-sheets, and concludes: "Hydrogen bonds play an *important* part in determining the configurations of these molecules."

Following the works of Pauling and his collaborators, the role of HBs in stabilizing the native form of proteins became well established. The HBs, with bond energies of the order of 24 kJ/mol, which provided an explanation for many anomalous properties of water, also took over the main cohesive forces needed for the organization of native proteins. In 1939, Pauling concluded:

I believe that as the methods of structural chemistry are further applied to the physiological problems, it will be found that the significance of the hydrogen bond for physiology is greater than that of any other single structural feature.

The first blow to the HB dominance came from the realization that though the HB *energy* is of the order of 24 kJ/mol, its formation in aqueous solution must have a negligible effect on the "driving force" for the process of protein folding. The argument apparently started with the work of Schellman (1955), summarized by Kauzmann (1959), and eventually was encapsulated in Fersht's HB inventory argument. It is seemingly straightforward and convincing. As we will soon see, it is also fundamentally wrong. The main argument is the following: write the stoichiometric reaction between a donor and an acceptor of an HB in the form

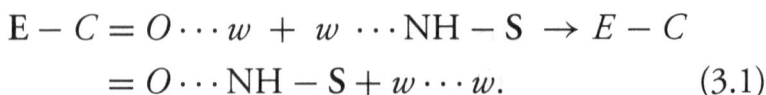

$$E - C = O \cdots w + w \cdots NH - S \rightarrow E - C$$
$$= O \cdots NH - S + w \cdots w. \qquad (3.1)$$

Here, E is an enzyme and S a substrate, but these could be two groups on the same protein. The two groups C=O and NH can belong to the same molecule or to two different molecules. This stoichiometric equation is well defined and meaningful. The main question is whether it is relevant to protein folding.

In 1955, John Schellman published an important article titled "The Thermodynamics of Urea Solutions and the Heat of Formation of the Peptide Bond." He was interested in the association of urea molecules. The urea molecules can form HBs between themselves, as well as with water molecules. The urea molecule (Figure 1.1) has a carbonyl group and an amine group. The first can serve as an *acceptor* for hydrogen bonding; the second can serve as a *donor* for hydrogen bonding, (Figure 1.2).

The very writing of the stoichiometric reaction by Schellman has led, on one hand to the dismissal of the contribution of HBs to the stability of proteins, and on the other hand, it led Kauzmann to propose the "hydrophobic bond" as a major factor in the stability of proteins. This will be discussed in Chapter 4.

Schellman considered various aggregations of urea mole-cules in water. For the purpose of this chapter, we focus only on the formation of dimers, i.e. one urea molecule contributes a donor and a second urea molecule contributes an acceptor to form an *inter*molecule HB.

Schellman concluded his article saying:

> *The peptide hydrogen bond, which is so important in modern theories of protein structure, involves groups which are almost identical with those of the urea molecules.*

This statement is, of course, correct. Indeed, the HBs formed by urea molecules are almost the same as the HBs formed by C=O and NH of the protein backbone.

It is also true that the HBs formed by urea molecules with water molecules have the same energy as the HBs between the C=O and NH with water molecules. However, the very writing of the stoichiometric reaction (3.1) suggests that when HBs between two molecules are formed in water, there is some kind of cancelation effect, and the two solute molecules are supposed to be initially hydrogen-bonded to water. When they form an intermolecular HB, they have to "sacrifice" the hydrogen bonding to water. In addition, the released water molecules can now form a new HB between themselves.

In 1959, Kauzmann summarized Schellman's experimen-tal results as follows (page 37):

> *From the rather small value of the f_u part of the formation of the hydrogen bond between urea molecules, J. A. Schellman (1955b) concluded that 'hydrogen bonds, taken by themselves, give a marginal stability to ordered structures, which may be enhanced or disrupted by interactions of side-chains.'*

(Here, f_u is referred to the "unitary" or the standard free energy of the reaction as depicted in [Figure 1.4(a)]).

Kauzmann inferred from the urea–urea hydrogen bonding that when an *intra*molecular HB is formed in protein the same argument of cancelation applies, namely that HBs cannot contribute much to the stability of the 3D structure of proteins. It should be said that Schellman himself denied that he reached the conclusion as quoted above from Kauzmann's article (Schellman, private communication). Schellman told me that he had always believed that HBs are important in the stability of the protein. It seems that Kauzmann arrived at his conclusion from Eq. 3.1 as it appeared in Schellman's article, and not from any conclusion made by Schellman himself.

3.3 The Evolution of the Myth that HBs Are Insignificant in Protein Stability

Following Schellman's experiments and Kauzmann's conclusion based on those experiments, it was widely believed that HBs could not contribute significantly to protein folding and protein stability.

If HBs are not the main factor in the stabilization of the structure of the protein, then what is that factor?

To answer this question, Kauzmann suggested that the "hydrophobic bond" could be that factor. We will discuss the hydrophobic "bond" or the hydrophobic effect in the next chapter. Here, we focus on the "stoichiometric" reaction as written in Figure 3.1. The conclusion reached by Kauzmann has been widely cited in the literature, including many textbooks of biochemistry. We present here a few quotations from recent textbooks:

> *The importance of the hydrophobic effect was for a long time underestimated. Charged interactions and hydrogen bonds are not strong intramolecular forces because water molecules compete significantly with these effects."*
> — Whitford (2005)

> *An important non-covalent force that causes a polypeptide to fold into its native conformation are the hydrophobic interaction forces. . . . Hydrogen bonds . . . although they contribute to the thermodynamic stability of protein's conformation, their formation may not be a major driving force for folding.*
> — Devlin (2006)

> *The hydrophobic effect, which causes non-polar substances to minimize their contacts with water, is the major determinant of native protein structure. . . . Perhaps surprisingly, hydrogen bonds, which are central features of protein's structures, make only minor contributions to protein stability.*
> — Voet *et al.* (2008)

Similarly, for the stability of DNA, they write (page 837):

> *It is clear that hydrogen bonding is required for the specificity of base pairing in DNA. Yet, as is also true for proteins . . . hydrogen bonding contributes little to the stability of nucleic acid structures. This is because, on denaturation, the hydrogen bonds between the base pairs of a native nucleic acid are replaced by energetically similar hydrogen bonds between the bases and water.*

I have seen many quotations of this argument, in textbooks, articles and lectures. Unfortunately, it is totally wrong. Because of its "importance", it was eventually given a name: "The HB inventory argument." This very impressive name has become a theorem, a paradigm or a dogma lending further credibility to its validity. Referring to the stoichiometric reaction as shown in [Figure 3.4(c)], Fersht (1999) writes:

> On the right, –XH bonds with B, and the released water molecules each make a hydrogen bond with water, which is formally equivalent to binding to each other.
>
> A crude, but very effective, way of understanding the energetics of hydrogen bonding is to perform a hydrogen bond inventory, i.e., count the number and nature of hydrogen bonds on each side of the chemical equation. Suppose that –XH is a good donor and B a good receptor. Then, on each side of the equation, there are two good hydrogen bonds. There is then little change in enthalpy in the reaction. Suppose that –XH is a poor hydrogen bond donor, but B is a good acceptor. Then on the left there is one strong bond and one weak one. The same is true on the right. The same inventory can be performed if –XH is a good donor and B is a poor acceptor. Again, there should be little change in enthalpy during the reaction. Thus, hydrogen bonding should be relatively isoenthalpic.
>
> Although the hydrogen bond inventory is zero, hydrogen bonding is energetically favorable in the formation of enzyme–substrate complexes because of the increase in entropy on the release of bound water molecules.

Unfortunately, this seemingly logical argument is totally wrong. I will explain why in the next section.

3.4 Why the Hydrogen Bond Inventory Argument is Wrong

The HB inventory argument is based on the following stoichiometric "reaction":

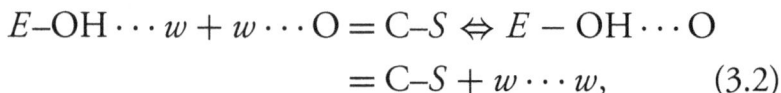

$$E\text{–}OH \cdots w + w \cdots O = C\text{–}S \Leftrightarrow E - OH \cdots O$$
$$= C\text{–}S + w \cdots w, \qquad (3.2)$$

where E and S may be an enzyme and a substrate; or any two FGs, one donor and one acceptor for hydrogen bonding. As we have noted earlier, perhaps the first "reaction" of the form (3.2) was written by Schellman (1955) in his study of the association between the urea molecules in water.

The term "HB inventory" was coined by Fersht (1985, 1987, 1999). Basically, it states that on the LHS of Eq. (3.2) a donor (E) makes an HB with a water molecule, and the acceptor (S) makes an HB with a water molecule. On the RHS, the acceptor and donor make an HB, and the released water molecules each make an HB with water, which is formally equivalent to binding to each other.

In another publication on molecular recognition, Fersht (1987) writes:

The inventory shows that the net change in the number of hydrogen bonds is zero.

It is not! This wrong conclusion follows from the very writing of the stoichiometric reaction (3.2) and then *counting* the number of HBs on each side of the equation. Unfortunately, the inference based on this equation is invalid. Furthermore, the *counting* involves different things on the two sides of the equation. Fersht continues his argument, focusing on the enthalpy change in this reaction, and concludes:

> *Thus, hydrogen bonding should be relatively isoenthalpic...*
> *although the hydrogen bond inventory is zero, hydrogen bonding*
> *is energetically favorable in the formation of enzyme–substrate*
> *complexes because of the increase in entropy on the release of bound*
> *water molecules.*

Similarly, Creighton (1993) writes:

> *Hydrogen bonds within proteins have not been considered recently*
> *to contribute to the net stability of the folded state since the unfolded*
> *protein was considered to make comparable hydrogen bonds to*
> *water.*

In my opinion, neither of the conclusions about the enthalpy, or about the entropy change in the reaction (3.2), is valid. First, because the "counting" of HBs on the two sides of Eq. 3.2 is not valid (see below), and second, the entropy change in this process cannot be predicted from just *looking* at the reaction (3.2) and observing the released water molecules. The entropy change associated with the process (3.2) could be positive or negative, depending on the "structural changes" in the water induced by the formation of an HB between E and S.

In this section, we will first discuss the validity of the HB inventory argument based on the very writing of a stoichiometric reaction of the form (3.2). Then, in Section 3.5, we will comment on the entropy and the enthalpy changes in this process.

Without specifying the thermodynamic quantity associated with this reaction, one can *see* from Eq. (3.2) that in the process of formation of an HB between E and S, two HBs are *broken* and two HBs are *formed*. Assuming that the strength of the HB energy is approximately the same between a hydroxyl and water, between a carbonyl and water and between two water molecules, one can reach the conclusion that the formation of an HB (inter- or intramolecular) will have a minor effect on the energetics of such a reaction.

Indeed, this conclusion is correct for the specific reaction as written in Eq. (3.2). However, this is not the reaction that takes place when an HB is formed between E and S. The correct reaction is not (3.2) but

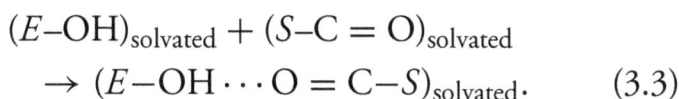

$$(E-OH)_{solvated} + (S-C = O)_{solvated}$$
$$\rightarrow (E-OH \cdots O = C-S)_{solvated}. \qquad (3.3)$$

Thus, an enzyme E and a substrate S form a complex ES with a direct HB. All these three components are *solvated* by water. To see that Eq. (3.3) is the correct equation, suppose that we do the same reaction, i.e. binding two urea molecules in the gaseous phase and in the liquid phase (Figure 3.3). We see that in the gaseous phase a genuine HB is formed between the two molecules. Once we transfer all the reactants and

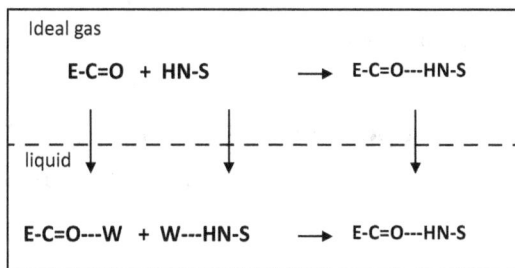

Fig. 3.3. Hydrogen bond formation in the gaseous and the liquid phases.

products of the reaction into the solvent, we get the reaction as in Eq. (3.3).

The replacement of the (correct) reaction (3.3) by the (incorrect) reaction (3.2) reveals the two flaws which invalidate the HB inventory argument.

First, the $H\phi I$ groups on the LHS of Eq. (3.3) do not *form* an HB with a water molecule but are *solvated* by water molecules. The HB formed by E and S on the RHS is a *genuine* HB. Thus, the mere counting of "bonds" on the two side of Eq. (3.2) is like counting apples and bananas on the two sides of the equation.

Second, the water molecules "released" in the reaction (3.2) are at the *same chemical potential* before and after the association of E and S. Therefore, the effect of the released water molecules is taken into account in the *difference* in the solvation Gibbs energies of the species E, S and ES. Both the enthalpy and the entropy changes associated with this association cannot be predicted by mere inspection of the reaction (3.2). These quantities depend on the "structural changes" in the water induced by the reaction. These structural changes can contribute positively or negatively to the entropy and the enthalpy of the reaction, but because of the exact entropy–enthalpy compensation they will have no effect on the Gibbs energy of the reaction.

Another way of arguing is to say that the water molecules which are released upon desolvation of the NH and C=O groups into the solvent "flow" at a constant chemical potential of the water in the solution.

Thus, when considering the energetics of a reaction involving the formation of an HB (either inter- or intramolecular) in aqueous solutions, one should count how many solvated "arms" of the $H\phi I$ groups are desolvated. For instance, an NH group has one arm along which an HB can be formed. A C=O has two arms and an OH has three arms (a water molecule has four arms, along which it can form an HB with other water molecules).

The solvation Gibbs energy of one arm was estimated, both theoretically and from experimental data [see Ben-Naim (2009, 2013)], to be about 2.25 kcal/mol.

The (genuine) HB energy is of the order to 6–6.5 kcal/mol. Therefore, when one arm forms an HB with another arm, the *loss* of solvation Gibbs energy is about 4.5 kcal/mol. The formation of one HB involves the *gain* of about 6–6.5 kcal/mol. The net change in Gibbs energy for such a reaction is therefore about 1.5–2 kcal/mol. This is not a negligible contribution to the stability of protein, as has been erroneously concluded for many years [see Ben-Naim (1991b)].

This analysis was carried out by Ben-Naim (1990b). It was concluded that direct intramolecular HBs could contribute *significantly* to the stability of the protein. The exact contribution depends on the number of *arms* of the $H\phi I$ groups which are desolvated. In addition, other $H\phi I$ effects were discovered. These are discussed in Chapter 4.

Unfortunately, in spite of the fact that the HB inventory argument was proven to be wrong [Ben-Naim (1991b)], it was repeated by Fersht (1999) and many others.

3.5 Another Way to Analyze the Gibbs Energy Change in the Formation of an HB

In Figure 3.3, we show the reaction between two urea molecules in solution. Similarly, for the reaction between E and S in Eq. (3.1), we can write

$$\Delta G(\text{reaction 3.2}) = \mu_{ES}^* - \mu_S^* - \mu_E^*, \qquad (3.4)$$

where the asterisk stands for the pseudo-chemical potential [Ben-Naim (2006)]; see also Appendix A. The solvation of the species ES, S and E includes the coupling work of all the species involved in the chemical reaction. For a very dilute system with respect to all of the species E, S and ES, we can write Eq. (3.4) in more detail (assuming for simplicity that the internal degrees of freedom are not affected by the solvation process):

$$\Delta G = W(ES|w) - W(S|w) - W(E|w)$$
$$+ k_B T \ln \frac{\rho_{ES}\rho_E\rho_S\Lambda_{ES}^3\Lambda_E^3\Lambda_S^3}{q_{ES}q_E q_S}. \qquad (3.5)$$

The solvent effect on ΔG of the reaction (3.2) can be written as the difference in the solvation Gibbs energies of the molecules involved:

$$\delta G = \Delta G^l - \Delta G^g = \Delta\mu_{ES}^* - \Delta\mu_E^* - \Delta\mu_S^*. \qquad (3.6)$$

Here, $W(\alpha|w)$ is the coupling work of α to the solvent w [Ben-Naim (2006)], Λ_α^3 the momentum partition function of α and q_α the internal partition function of the species α.

Thus, the reaction that actually occurs is (3.3) and all the thermodynamic quantities associated with this reaction should be derived from the Gibbs energy change (3.4).

Note that in examining the stoichiometric reaction as in Eq. (3.2), one must realize that the ellipsis, denoting an HB, does not have the same meaning on the RHS and LHS of this reaction. Whereas the hydroxyl and carbonyl groups are *solvated* by water molecules on the LHS of (3.2), there is a *genuine* HB between the hydroxyl and the carbonyl group on the RHS of (3.2). However, if one insists on examining stoichiometric reactions of the type (3.2), one must consider many possible "reactions" of this type. A few examples of these are shown in Figure 3.4 (here the ellipses denote genuine HBs).

Clearly, there are many stoichiometric reactions of the form (3.2). In some, the hydroxyl group may be engaged in zero, one, two or three HBs with water molecules. Similarly, the carbonyl group may be engaged in zero, one or two HBs. On the RHS of the reaction, an HB is formed by E and S. The water molecules released may or may not form an HB between themselves. Each of these reactions has a well-defined standard Gibbs energy.

Thus, if one prefers a specific stoichiometric reaction of the type (3.2), one should realize that this reaction does not represent what *actually happens* when an HB is formed between E and S, but this is only one out of the many possible reactions of the type shown in Eq. (3.2). The simpler approach is, of course, to study the reaction (3.3), which takes into

E-OH·····w + S-C=O·····w ⟶ E-OH·····O=O-C + w·····w

E-OH·····w + S-C=O·····w ⟶ E-OH·····O=O-C + w + w

E-OH⟨w + S-C=O·····w ⟶ E-OH·····O=O-C + w·····w + w

E-OH⟨w,w + S-C=O·····w ⟶ E-OH·····O=O-C + w·····w + 2 w

E-OH·····w + S-C=O⟨w,w ⟶ E-OH·····O=O-C + w·····w + w

E-OH⟨w,w + S-C=O⟨w,w ⟶ E-OH·····O=O-C + w·····w + 2 w

E-OH⟨w,w + S-C=O⟨w,w ⟶ E-OH·····O=O-C + w·····w + w·····w

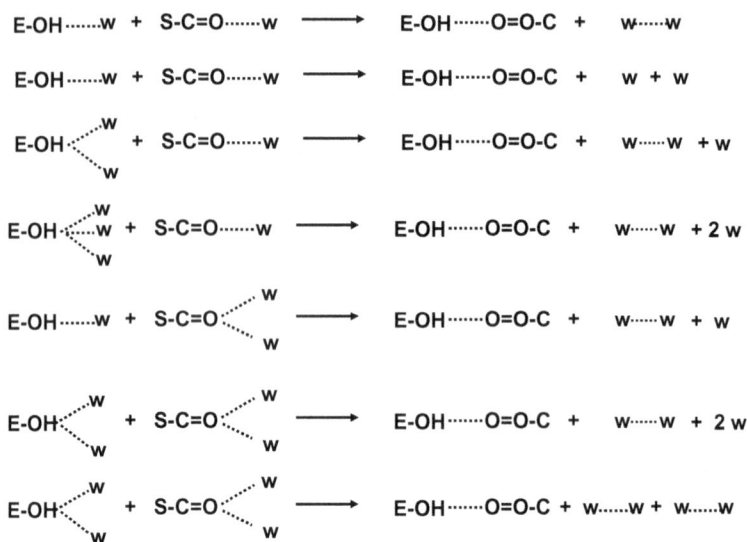

Fig. 3.4. Various stoichiometric reactions involving direct hydrogen bonding.

account all the possibilities of the reactions in Figure 3.4 through the *solvation* Gibbs energies of the species E, S and ES.

Note that Figure 3.4 lists only a very few of the possible reactions involving *real* HBs. In fact, one must also take the reaction between E and S, which interact with water molecules through weak interactions. These would make the list in the Figure 3.4 infinitely long.

To summarize, the stoichiometric reaction as written in (3.2) might lead to erroneous conclusions. First, because the HB formed on the RHS is a genuine HB and contributes an HB energy. On the other hand, the ellipses on the LHS do not represent a genuine HB but a solvation of a hydroxyl and a carbonyl group by water. Second, the water molecules released into the solvent are the same molecules that solvate the E and S molecules; they have the same chemical potential and therefore these molecules do not contribute a genuine HB energy, as might be implied by the very writing of the reaction (3.2). Estimates of the actual contribution of an HB in aqueous solutions were made depending on the number of "arms" which became buried in the interior of the protein [Ben-Naim, (1991b)].

3.6 The Entropy and the Enthalpy Change Associated with the Reaction (3.3)

Some authors discuss the effect of the release of the water molecules on the entropy or the enthalpy change in the process. This is a potentially interesting problem. However, it is very difficult, if not impossible, to reach a quantitative estimate of the contribution of the released water molecules to the entropy and the enthalpy change in a reaction such as (3.3). Furthermore, these changes are irrelevant to the Gibbs energy change of the process. We present below a simple argument based on the mixture model approach.

The exact entropy–enthalpy compensation was first shown in a two-structure model for water, and later it was proven in the general form [Ben-Naim (2009)]. Here, we use a simple mixture model for water to show that the water release in the process (3.3) may affect the entropy and the enthalpy of the process, but cannot affect the Gibbs (or Helmholtz) energy of the process.

Consider the following mixture model view of liquid water [Ben-Naim (2009)]. For each snapshot of the system we can classify all water molecules into five groups, and count the number of molecules in each group. The vector $(N_0, N_1, N_2, N_3, N_4)$ describes the composition of the mixture of species at a specific configuration of the system, where N_i is the number of water molecules engaged in i HBs. We can also define the *average* composition of the system by the vector $(\overline{N}_0, \overline{N}_1, \overline{N}_2, \overline{N}_3, \overline{N}_4)$. With this vector, one can define the concept of the structure of water, but we will not need this concept here. Whenever a process occurs in aqueous solutions, there might be change in the composition of the system (as well as in the *structure* of water).

Within the mixture model approach to water we can write the chemical potential, the partial molar entropy and the partial molar enthalpy of any solute A as

$$\mu_A = \left(\frac{\partial G}{\partial N_A}\right)_{N_w} = \left(\frac{\partial G}{\partial N_A}\right)_{\overline{N}_0,\ldots,\overline{N}_4}$$

$$+ \sum_{i=0}^{4} \left(\frac{\partial G}{\partial \overline{N}_i}\right)\left(\frac{\partial \overline{N}_i}{\partial N_A}\right), \qquad (3.7)$$

$$\overline{S}_A = \left(\frac{\partial S}{\partial N_A}\right)_{N_w} = \left(\frac{\partial S}{\partial N_A}\right)_{\overline{N}_0,\ldots,\vec{N}_4}$$

$$+ \sum_{i=0}^{4} \left(\frac{\partial S}{\partial \overline{N}_i}\right)\left(\frac{\partial \overline{N}_i}{\partial N_A}\right)_{N_w}, \qquad (3.8)$$

$$\bar{H}_A = \left(\frac{\partial H}{\partial N_A}\right)_{N_w} = \left(\frac{\partial H}{\partial N_A}\right)_{\vec{N}_0,\ldots,\vec{N}_4}$$

$$+ \sum_{i=0}^{4} \left(\frac{\partial H}{\partial \bar{N}_i}\right)\left(\frac{\partial \bar{N}_i}{\partial N_A}\right)_{N_w}. \qquad (3.9)$$

All the derivatives are at constant P, T and the number of any other solute molecules. The derivative at constant N_w means that the *total* number of water molecules, $N_w = \overline{N}_0 + \overline{N}_1 + \overline{N}_2 + \overline{N}_3 + \overline{N}_4)$, is kept constant, but each of the $\overline{N}_i$ may change. On the other hand, a derivative at constants $\overline{N}_0, \ldots, \overline{N}_4$ means that we add N_A molecules of A in a hypothetical liquid where the conversion between all the species of water and molecules is "frozen-in."

Each of the partial molar quantity has two terms. First, we add N_A in a frozen-in (fr) system, and then we release the constraint on fixed values of $\overline{N}_0, \ldots, \overline{N}_4$ and let the system reach a new equilibrium composition. We may refer to the first term as the frozen-in (fr) partial molar quantity, and the second term as the relaxation term, or as the *structural* changes in the solvent induced by the process of adding N_A. We rewrite Eqs. (3.7)–(3.9) as

$$\mu_A = \mu_A^{\text{fr}} + \sum_{i=0}^{4} \mu_i \left(\frac{\partial \overline{N}_i}{\partial N_A} \right)_{N_w}, \qquad (3.10)$$

$$\overline{S}_A = S_A^{\text{fr}} + \sum_{i=0}^{4} \overline{S}_i \left(\frac{\partial \overline{N}_i}{\partial N_A} \right)_{N_w}, \qquad (3.11)$$

$$\overline{H}_A = H_A^{\text{fr}} + \sum_{i=0}^{4} \overline{H}_i \left(\frac{\partial \overline{N}_i}{\partial N_A} \right)_{N_w}. \qquad (3.12)$$

Note that the bars on $\overline{S}_A$ and $\overline{H}_A$ are for the partial molar quantity, whereas the bar on $\overline{N}_i$ is for the average number of the species i.

At equilibrium, we must have the equality of the chemical potentials

$$\mu_w = \mu_0 = \mu_1 = \mu_2 = \mu_3 = \mu_4, \qquad (3.13)$$

where μ_w is the chemical potential of the water and μ_i is the chemical potential of the species i of water molecules.

Using the equilibrium condition (3.13), we can rewrite the second term on the RHS of (3.10) as

$$\sum_{i=0}^{4} \mu_i \left(\frac{\partial \bar{N}_i}{\partial N_A} \right)_{N_w} = \mu_w \sum_{i=0}^{4} \mu_i \left(\frac{\partial \bar{N}_i}{\partial N_A} \right)_{N_w}$$

$$= \mu_w \frac{\partial}{\partial N_A} \left(\sum_{i=0}^{4} \bar{N}_i \right) = 0. \qquad (3.14)$$

The last equality follows from the fact that $\sum_{i=0}^{4} \bar{N}_i$ is constant.

Thus, we can rewrite Eqs. (3.10)–(3.12) as

$$\mu_A = \mu_A^{\text{fr}}, \qquad (3.15)$$

$$\bar{S}_A = S_A^{\text{fr}} + \Delta S_A^{\text{st}}, \qquad (3.16)$$

$$\bar{H}_A = H_A^{\text{fr}} + \Delta H_A^{\text{st}}. \qquad (3.17)$$

This result is important. As we have shown in Chapter 2, structural changes in the solvent can affect the partial molar entropy, enthalpy, volume, etc. of a solute A, but not the chemical potential of water. The reason for this result is the chemical equilibrium condition (3.13). When we add dN_A molecules of A, the "flow" of molecules from one species to another is at a constant chemical potential of water molecules.

The last result can be applied to any process occurring in water. For instance, for the isomerization reaction $A \rightarrow B$, we have

$$\Delta G(A \rightarrow B) = \mu_B - \mu_A = \mu_B^{\text{fr}} - \mu_A^{\text{fr}} = \Delta \mu^{\text{fr}}, \quad (3.18)$$

$$\Delta S(A \rightarrow B) = \overline{S}_B - \overline{S}_A = \Delta S^{\text{fr}} + \Delta S_B^{\text{st}} - \Delta S_A^{\text{st}}, \quad (3.19)$$

$$\Delta H(A \rightarrow B) = \overline{H}_B - \overline{H}_A = \Delta S^{\text{fr}} + \Delta H_B^{\text{st}} - \Delta H_A^{\text{st}}. \quad (3.20)$$

Thus, in any process occurring in water, there can be a redistribution in the composition of the water, which we refer to as "structural changes" in water. These changes may affect the entropy, enthalpy, volume, etc. of that reaction. However, the change in Gibbs energy (or the "driving force") for the process is not affected.

Specifically, for the dimerization of urea in water, or the intramolecular hydrogen bonding in protein, some water molecules are released from the solvation sphere of the two functional groups that form HBs. However, this release of the water molecules occurs at a constant chemical potential of the water. Therefore, the water molecules which are released in the process cannot affect the driving force for the process.

Finally, it should be noted that neither the sign nor the magnitude of the quantities ΔS_a^{st} may be easily estimated. A process can cause a redistribution of the "composition" of the solvent (any solvent) when viewed as a mixture of species. This change of composition will cause changes in the entropy, enthalpy, volume, etc. of the system. To predict the sign of the changes in these thermodynamic quantities, one must know the partial molar quantities of all the species involved, as well as the changes in the average number of the molecules of each species. An exact argument based on the Kirkwood–Buff theory of solution for the case of solvation of inert gas molecules in water is provided in Ben–Naim (2009).

3.7 Conclusion

We have seen that the so-called HB inventory argument, which is based on a stoichiometric reaction of the form (3.1) is invalid. The pitfall is the very writing of such a reaction and counting different things on the two sides of the equation. In the historical perspective, this argument has led to dismissing the contribution of hydrogen bonding in protein folding and to the birth of the $H\phi O$ effect, which reign supreme in the field of protein folding. Not only were *direct* HBs deemed to contribute insignificantly to the stability of proteins, but the resulting holding onto the $H\phi O$ effect was so tight and overwhelming that most people in the field did not even search for other possible solvent-induced effects. In particular, no one had even suggested that other effects involving $H\phi I$ groups might contribute significantly to the stability of the protein. Thus, the HB inventory argument was not only the cause of the birth of the $H\phi O$ effect, but it also fueled the evolving myth that the $H\phi O$ effect is the dominant effect in protein folding. This dogma has survived even to this day. We will discuss this myth in the next chapter. The conclusion reached by Kauzmann based on Schellman's experiments, that hydrogen bonding does not contribute significantly to the stability of proteins, is only a myth.

4

From Kauzmann's Conjecture to the Myth That the Hydrophobic Effect is the Dominant Factor in Protein Stability

4.1 Abstract

As noted in the previous chapter, the rise of the $H\phi O$ effect is intimately associated with the fall of the hydrogen bonding in protein folding and protein stabilization. In this chapter, we will discuss the evolution of Kauzmann's cautionary conjecture on the role of the $H\phi O$ effect into becoming the "most important" and the "dominant" effect in protein folding, protein–protein association and molecular recognition. In short, the $H\phi O$ effect became a panacea for all problems in molecular biology.

That the $H\phi O$ effect is dominant in protein folding is perhaps the most widespread myth. It is also understandable why this myth was almost universally accepted until the late 1980s. I myself believed the solvation of nonpolar molecules to be important in protein folding. In the early 1970s, I had

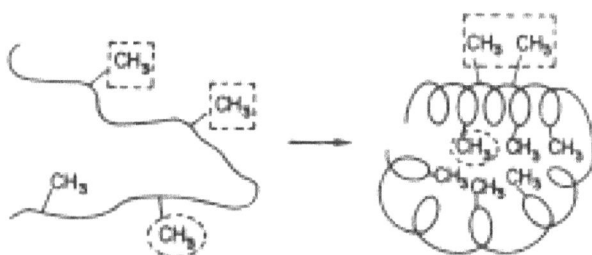

Fig. 4.1. Schematic "extraction" of hydrophobic "solvation" and "interaction" between hydrophobic groups.

dedicated much of my research efforts to both the definition and the measurability of the pairwise $H\phi O$ interaction. I also believed that these interactions were important for protein folding. Figure 4.1 reproduces the first figure from my book (1980) where the solvation and pairwise interaction between $H\phi O$ groups were "extracted" from the PFP to be studied in isolation. It was only in 1989 that I realized that both the solvation of $H\phi O$ molecules and the pairwise potential of mean force (PMF) between $H\phi O$ molecules are not only not dominant in protein folding, but in fact are *irrelevant* to protein folding. Instead, the solvation, the HBs and the pairwise PMF between hydrophilic groups are far more important. In this chapter, we will review all the available "evidence" in favor of the $H\phi O$ effect, as well as the evidence in favor of the $H\phi I$ effect in protein folding.

In spite of the fact that the overwhelming evidence is in favor of the $H\phi I$ effect in protein folding, protein–protein association and molecular recognition, as were published over 20 years ago, people continued to cling to the old myth of the dominance of the $H\phi O$ effect. (Why this is the case is an interesting question which I will discuss in the Epilogue.)

4.2 The Origin of the Myth and Some Historical Notes

Open any textbook of biochemistry, biophysics or molecular biology, look at the index for the entry "hydrophobic effect," and you are likely to find a statement claiming that the hydrophobic effect (or bond, or interaction) is the *most important* driving force in some biochemical processes such as protein folding and protein–protein association. The origin of these ideas may be found in Kauzmann's original article (1959).

As was explained in Chapter 3, the so-called HB inventory argument in effect dismissed the contribution of HBs to protein folding. If HBs do not provide the main "driving force" for protein folding, what factors do provide those "driving forces"? This apparent conceptual vacuum was filled by the $H\phi O$ effect in 1959. The $H\phi O$ effect was known long before Kauzmann applied it to the problem of protein folding. It was applied successfully to explain the surface tension of certain aqueous solutions of organic molecules, micelle formation and membranes. All these phenomena involve molecules having two moieties: a hydrophobic part, which "fears" water and tries to avoid it, and a part which "loves" water and mingles with it comfortably. As Tanford and Reynolds (2001) quoted from a personal communication with Kauzmann, the idea of the $H\phi O$ effect had been hovering "in the air" for a long time.

In his review article "Some Factors in the Interpretation of Protein Denaturation," Kauzmann (1959) applied the idea of the $H\phi O$ effect to protein folding. For this purpose he coined the term "$H\phi O$ bond," and speculated that this "bond" could be a major factor in the stabilization of the native structure of protein. Here is the quotation from that article:

> Hydrogen bonds, taken by themselves, give a marginal stability to ordered structures.... The hydrophobic bond is probably one of the more important factors involved in stabilizing the folded configuration in many native proteins.

It should be noted that the idea that the *solvent* might play a major role in biochemical processes occurring in aqueous solution was expressed by Kirkwood (1954) at a conference on "Mechanism of Enzyme Action." However, at that time it was not clear how the solvent could contribute to the stability of protein structure.

Kauzmann's idea was very simple and convincing. It was known that the Gibbs energy of transferring a small nonpolar solute such as methane or ethane from water into an organic liquid involves a large negative change in the Gibbs energy. This is the same driving force for the formation of micelles and membranes in aqueous solutions. Kauzmann also noticed that about one-third of the side chains of a typical protein are $H\phi O$, and most of these find themselves buried in the interior of the folded protein. If we can take the process of transfer, as shown in [Figure 4.2(a)], to represent the process of transfer of the side chain as shown in [Figure 4.2(b)], then we can estimate that a protein of about 150 amino acids has about 50 $H\phi O$ groups, and if each of these contributes between -12 and $-16\,kJ/mol$, we get a very large driving force for the folding process.

Kauzmann's idea was bold and brilliant and had captured the imagination of many scientists. It is not surprising,

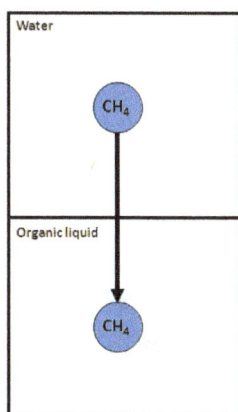

(a)

Transfer of a hydrophobic molecule from water into an organic liquid

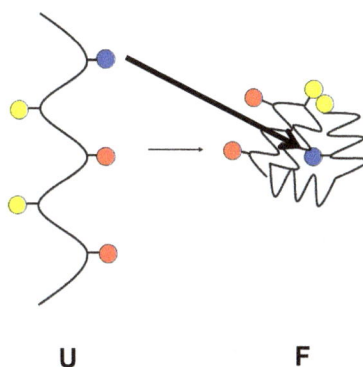

(b)

Transfer of a hydrophobic Group (blue) from water into the interior of the protein

Fig. 4.2. Kauzmann's model for the hydrophobic effect.

therefore, that the dominance of the $H\phi O$ effect has prevailed for over half a century. The fact that $H\phi O$ groups are in the *interior* of the *protein*, and the fact that the transfer of $H\phi O$ molecules from water to an organic liquid is large and negative, are undeniable. The former lends credibility to Kauzmann's model, while the latter provides the large negative Gibbs energy change.

It is not surprising that the use of the $H\phi O$ effect has pervaded the literature explaining many biochemical processes, such as protein folding, protein–protein association and molecular recognition. Very few have questioned the validity of Kauzmann's model for the process of protein folding, or the validity of its explanation based on Frank and Evans' ideas of iceberg formation (see Chapter 3). The $H\phi O$ paradigm was seemingly unassailable, and became the central dogma in the biochemical literature. This was understandable. There was no other solvent-induced effect which could compete with the $H\phi O$ effect as the main factor in the stability of protein.

In the early 1970s, I suggested to Kauzmann to replace the term "hydrophobic bond" and to make a clear cut distinction between different $H\phi O$ effects: on one hand the solvation of $H\phi O$ molecules, and on the other hand the interaction between two or more $H\phi O$ solutes. Note that at that time there was no single measure of the *strength* of the "pure hydrophobic" interaction.

In fact, Kauzmann's description of the $H\phi O$ bond is quite vague:

> *This tendency of the non-polar groups of proteins to adhere to one another in aqueous environments had been referred to as* **hydrophobic bonding**. *The hydrophobic bond is probably one of the most important factors involved in stabilizing the folded configuration in many native proteins. The most appropriate model systems for this discussion are solutions of hydrocarbons in water.*

It is clear that Kauzmann was talking about $H\phi O$ bonds, i.e. $H\phi O$ *interactions*. However, he suggested to study the *solvation* of $H\phi O$ *molecules*. It was only in the early 1970s that $H\phi O$ *interactions* were defined and several measures of the $H\phi O$ interaction were suggested. These are described in Ben-Naim (1980, 2009).

The formulation of the $H\phi O$ interactions in terms of the PMF, and the recognition of the irrelevance of the *direct* interaction to the study of the $H\phi O$ interaction, have led to a simple measure of the strength of the $H\phi O$

interaction between two methane molecules. For the first time, an experimental measure of the strength of the $H\phi O$ interactions was also suggested. An upsurge in publications on the theoretical aspects of $H\phi O$ interactions then ensued.

4.3 The Evolution of the Hydrophobic Effect in Protein Folding

If one reads carefully Kauzmann's conjecture, one finds that he was very cautious. He used the words "probably one of the more important" and "tentative conclusions." Here is a quotation from his main conclusion:

> From the above considerations we may draw the following some-what tentative conclusions about the thermodynamic properties of hydrophobic bonds involving the non-polar side chains found in proteins:
>
> 1. These bonds are stabilized largely by entropy effects. For each non-polar aliphatic side chain which leaves the aqueous environment and enters a non-polar region of the protein a gain in entropy of the order of 20 entropy units may be expected. This value may be more or less independent of the length of the side chain in alanine, valine and leucines, if we may extrapolate from the behavior of methane, propane and butane. For phenylalanine the entropy gain is probably half as great.
> 2. The transfer of an aliphatic side chain from water to a non-polar region in the protein is endothermic to the extent of 1000 to 2000 cal per mole of groups. For phenylalanine the transfer is more nearly athermal.
> 3. The free energy change in the transfer from water to a non-polar environment is exergonic to the extent of about 3000 to 5000 cal per mole of groups at room temperature.

Kauzmann was also aware of the fact that the solvation of a nonpolar group might be different from the solvation of the same group when it is attached to the backbone of the protein. From this cautionary conjecture, the $H\phi O$ effect became the "most important," "the dominant force" and the like. Here are a few quotations from recent books and articles:

> *An important non-covalent force that causes a polypeptide to fold into its native conformation are the hydrophobic interaction forces.... Hydrogen bonds... although they contribute to the thermodynamic stability of protein's confirmation, their formation may not be a major driving force for folding.*
> — Devlin (2006)

> *The hydrophobic effect, which causes non-polar substances to minimize their contacts with water, is the major determinant of native protein structure... hydrogen bonds, which are central features of protein's structures, make only minor contributions to protein stability.*
> — Voet *et al.* (2008)

> *Hydrophobic interactions are important in protein folding, denaturation... and packing. They are a major driving force for protein tertiary structure.*
> — Dias *et al.* (2011)

Most authors will not bother to present the evidence for such extraordinary statements. Instead, they will refer the reader to the relevant literature. Most readers will not bother to look at the relevant literature, and if they do they will find arguments that they cannot comprehend and will accept the textbook's statements on faith.

A few authors will present the "evidence" for the importance of the hydrophobic effect. Briefly, this evidence can be summarized in two sentences:

(i) The Gibbs energy of transferring nonpolar molecules from water to an organic liquid is known to be large and negative.

(ii) It is known that most nonpolar groups of the protein are found in the interior of the native structure.

These two experimental facts are more than enough to convince the reader that in the process of protein folding the "burial" of the hydrophobic groups in the interior of the protein provides the "driving force" for the process of folding. In Section 4.5, we will show why this argument is fallacious.

A few authors will also venture to "explain" the molecular origin of the hydrophobic effect. Citing the classical paper by Frank and Evans, they make the following two statements:

(iii) The Gibbs energy of transferring a nonpolar molecule from an organic liquid to water is "entropy-driven."

(iv) The large negative entropy change in the process of solvation of a nonpolar molecule in water is due to "ordering" or to "iccberg formation" around the nonpolar solute.

From the last two statements, one can easily conclude that the "ordering" of water molecules by a nonpolar molecule *explains* the large positive Gibbs energy of solvation, and hence also the source of the large "driving force" for the process of protein folding. This apparent "explanation" of the $H\phi O$ effect was discussed in Chapter 2.

In the rest of this chapter, we show that although statements (i), (ii) and (iii) are experimental facts, and statement (iv) may be justified theoretically, the conclusions derived from these statements are not warranted. This finding is sufficient to demolish the myth that the hydrophobic ($H\phi O$) effect is the most important factor in biochemical processes. It will also be shown that the $H\phi O$ effect as proposed by Kauzmann is *irrelevant* to the PFP. Instead, a new paradigm, based on a rich repertoire of hydrophilic ($H\phi I$) effects, has emerged. The $H\phi I$ effects provide both *strong interactions* and *strong forces*. The various $H\phi I$ *interactions* provide a sound explanation for the stability of the 3D structure of the proteins as well as the stability of aggregates of macromolecules. The $H\phi I$ *forces* will be discussed in Chapter 5 in connection with Levinthal's question. Of course, there is also some contribution due to the $H\phi O$ effects. However, these $H\phi O$ effects are neither the *solvation* nor the PMF associated with $H\phi O$ molecules, but the corresponding *conditional solvation* and *conditional* PMF. These are the real $H\phi O$ effects, but in reality they are relatively insignificant compared with the corresponding $H\phi I$ effects.

4.4 Cracks in the Kauzmann Conjecture About the Hydrophobic Effect

It was not until the 1980s that some doubts seeped in about the importance of the various $H\phi O$ effects to protein folding. First came some doubts about the importance of the $H\phi O$ effect in protein folding.

In 1980, in the preface to my book *Hydrophobic Interactions* [Ben-Naim (1980)], I wrote:

> *In spite of my researches in this field over almost 10 years, I cannot confirm that there is at present either theoretical or experimental evidence that unequivocally demonstrates the relative importance*

of the HφO interactions over other types of interactions in aqueous solutions.

My doubts were based on *lack of evidence* for the contention that the *HφO* effect is the *most important* effect in the driving force for protein folding. How can one claim that one factor is more important, or the most important, when one does not have a full *inventory* of *all* the factors involved in protein folding? Remember that Kauzmann's paper was on "*some factors* in the interpretation of protein denaturation" — not on *all factors* involved. Nobody knew what *all the factors* were, especially the solvent-induced ones. The only factor that could have competed with the *HφO* effect was the HB, but HBs were already deemed to be unimportant in aqueous solutions.

It should be emphasized here that my doubts were concerned with the contribution of various *HφO* effects to the stability of proteins. I discussed this topic several times with Kauzmann, who agreed that there is no strong evidence that the *HφO* is a dominant factor in the stability of proteins.

Kauzmann's model of inference from the transfer of molecules from water to organic liquid, and the fact that most *HφO* groups are found in the interior of the protein, were so convincing that the mere expression of doubts could not have threatened the dominance of the *HφO* dogma. One needs more than doubts. One needs facts! "Lack of evidence" for an idea cannot be used as evidence against that idea.

This was the main motivation for the examination of the entire question of the solvent-induced effects on protein folding and protein–protein association that I undertook late in the 1980s. (See next section.) The results of this examination were stunning — initially to me, then slowly diffusing into the literature.

First, there came the realization that the HB inventory argument was fundamentally faulty [Ben-Naim, (1991b)]. This was discussed in Chapter 3. Second, Kauzmann's model, appealing as it was, for over 50 years was found to be *irrelevant* to the protein folding process [Ben-Naim, (1990b)]. Finally, a logical pitfall: the fact that $H\phi O$ groups are found in the interior of the protein cannot be used as an argument in favor of the role of the $H\phi O$ effect in protein folding. Thus, by the end of the 1980s, my doubts about the role of $H\phi O$ effect in protein folding were not only validated but it was found that various hydrophilic effects are far more important.

This fact did not deter people from clinging to the myth that the $H\phi O$ effect is the most important factor in protein folding.

I will conclude this section with a quotation from Tanford and Reynolds' recent book (2001). This is an excellent book on the history of the field of protein chemistry. Nevertheless, I disagree with the authors' views as expressed in the quotation:

> *It is easy to underestimate the importance of the hydrophobic principle and to think that all that counts is the structure itself and that the question of whether one force or another dominates in forming the structure is a mere technical quibble, of the kind enjoyed by theoretical physical chemists. On the contrary, the matter is at the very heart of protein science. This can be illustrated by reference to a subject that is increasingly attracting attention, the problem of trying to predict three-dimensional structure on the basis of the amino acid sequence of a protein, plus general rules that govern polypeptide folding and internal organization.*

Harold Scheraga, professor at Cornwell University, was one of the first to try his hand at structure prediction, but he was also a great enthusiast for the hydrogen bond theory of proteins, one of the last to favour it. In 1960 he predicted a hypothetical three-dimensional structure for the protein ribonuclease, based on the primary sequence of its 124 amino acid residues, which had just recently been determined. The predicted structure, depicted in detail with atomic models, had the protein held together by internal hydrogen bonds between side-chain polar groups. This of course put many polar side chains in the centre of the structure, and left most of the hydrophobic groups dangling out at the surface. As it turned out, 1960 was also the year when the first high-resolution three-dimensional structure obtained by X-ray crystallography was published. In this, as in all the other protein structures published since then (including ribonuclease), the reverse situation prevails: charged groups and most polar side chains are at the surface, in contact with water, and most hydrophobic groups are "inside."

Could anything be more convincing? Scheraga made no stupid errors (of the sort made by Dorothy Wrinch); he constructed his helices exactly as Pauling prescribed, etc. He did everything right in his model building except in the assumption he made at the very start, as to the nature of the bonds and forces that dominate in making a protein what it is. The same principle would apply to theoretical analyses of functional properties (for example, substrate binding, transport in and out of the cell) — they would confuse and mislead if based on wrong assumptions. All interactions of proteins are governed by the same forces and cannot be understood even approximately without appreciating the thermodynamic dominance of hydrophobicity.

Of course, there are more examples of authors claiming that the $H\phi O$ effects dominate the process of protein folding. Before one can claim anything about the relative importance of one factor over another in maintaining the stability of proteins, one must have at least a complete inventory of all possible factors which contribute to the stability of proteins. The direct potential function for protein folding is relatively well known and well understood. The *indirect* solvent-induced effects are far more difficult to understand. Until the late 1980s, there was no inventory of all possible solvent-induced effects.

I undertook to establish such an inventory in the late 1980s. Here, I will discuss only two $H\phi O$ effects and two $H\phi I$ effects.

4.5 An Inventory of All Possible Solvent-induced Effects

In this section, we discuss very briefly the methodology of obtaining a complete inventory of all possible solvent-induced effect [Ben-Naim (1990b, 1992, 2011d)].

In Figure 4.3, we show schematically the process of protein folding. For simplicity, we assume that there is one representative of an $H\phi O$ side chain — say, a methyl group shown in light blue — and a representative of an $H\phi I$ group [either of the side chains or belonging to the backbone (BB)], such as C=O, NH or OH, shown as red circles.

We write the Gibbs energy change for the folding process as

$$\Delta G^{l}(\text{U} \rightarrow \text{F}) = \Delta U(\text{U} \rightarrow \text{F}) + \delta G(\text{U} \rightarrow \text{F}), \quad (4.1)$$

focusing on the solvent-induced effect δG, which we rewrite as

$$\delta G(\text{U} \rightarrow \text{F}) = \Delta G_{F}^{*} - \Delta G_{U}^{*}. \quad (4.2)$$

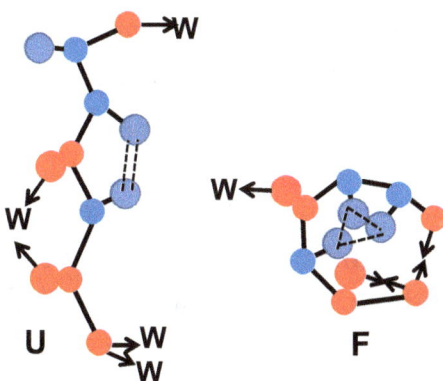

Fig. 4.3. A segment of a protein with a few hydrophobic groups (blue), and hydrophilic groups (red).

Before we continue, it should be said that we are discussing here the Gibbs energy change for the "reaction" as shown in Figure 4.3. Here, the U and F forms are represented by one conformation for each. We are interested in establishing an inventory of solvent-induced effects. For this purpose, it is sufficient to discuss only the process of conversion from *one* conformation, denoted U, to another conformation, denoted F. When discussing the experimental Gibbs energy of folding, one must take the average over all possible conformations belonging to the unfolded and the folded forms of the protein.

Thus, in Eq. (4.2), the solvation Gibbs energies ΔG_F^* and ΔG_U^* correspond to the solvation of one conformer of the F and U forms, respectively.

Next, we want to re-express the change in the solvation Gibbs energies in terms of the "fates" of all the functional groups which are exposed to the solvent. We will do the following analysis in a qualitative way. More mathematical analysis is available in Ben-Naim (1990b, 1992, 2011d).

Suppose that we cutoff all the functional groups (FG) in both the F and the U form. We first carry out the process of folding for the BB. The corresponding Gibbs energy change involved in this process is related to the *excluded volume* of the F and U molecules. Here, the "excluded volume" is with respect to solvent molecules. Next, we add all the FGs, but we do this in steps. First, we add all the FGs which are independently solvated. By "independently solvated" we mean that the conditional solvation of one FG given the BB is independent of the conditional solvation of the second FG given the BB. Next, we add all the pair-correlated FGs, then the triple-correlated, and so on.

We will not write here the formal expansion of each of the solvation Gibbs energies on the RHS of Eq. (4.2). Instead, we look at Figure 4.3 and record the changes in the fates of all the FGs upon the process of folding. We see that some of the $H\phi O$ groups which are initially solvated by water are found in the interior of the protein. This process involves the *conditional* solvation Gibbs energy of these groups in the U form, and the interaction energy of these groups with their surroundings in the F form. Similarly, an $H\phi I$ group is desolvated in the U form and interacts with its surroundings in the F form. This interaction mainly involves hydrogen bonding in the interior of the F form.

Next, we consider pair-correlated $H\phi O$ groups (shown as double-dashed lines in Figure 4.3). Again, we emphasize that the correlated $H\phi O$ involves the *conditional* PMF between two $H\phi O$ groups. Such pair correlations between two $H\phi O$ groups can occur either in the F or in the U form or in both. Similarly, we have to consider pair correlation between two hydrophilic groups (one such pair is shown in the U form in Figure 4.3).

In principle, one can continue with a triplet of FGs, a quadruplet of FGs, etc. However, we stop here after having considered the solvation of FGs and the pair correlation of FGs.

Some important conclusions are as follows.

The analysis of *all* the solvent-induced factors reveals that Kauzmann's model *does not* feature in the driving force for the process of protein folding. Instead of the Gibbs energy of *solvation* of an $H\phi O$ *molecule* in water, the *conditional solvation* Gibbs energy of an $H\phi O$ *group* features in the driving force. These Gibbs energies are very different from the Gibbs energies of solvation in water. The difference could be one or two orders of magnitude. The main reason is that an $H\phi O$ group attached to the BB of the protein is surrounded by water molecules which are perturbed by the BB (Figure 4.4). On the other hand, the solvation of an $H\phi O$ molecule is by unperturbed water molecules. Thus, at this

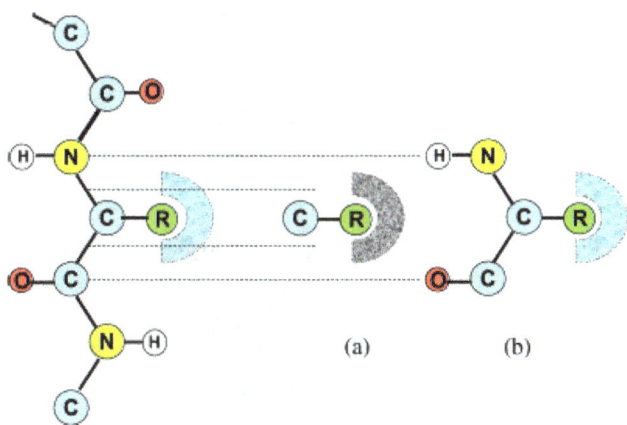

Fig. 4.4. The environment of a group R, when it is attached to a real protein and to a small segments of the protein.

stage, not only the basis on which the $H\phi O$ model was built was demolished, but the $H\phi O$ model itself was now shown to be inadequate. In the next section, we will also compare the magnitude of the $H\phi I$ and the $H\phi O$ effect. We will find that in each case the $H\phi I$ effect contributes much more than the corresponding $H\phi O$ effect.

It is unfortunate that despite these findings people still propagate the nonevidential "evidence" in favor of the $H\phi O$ effect. In the next two sections, we will discuss separately the evidence for the $H\phi O$ and the $H\phi I$ effect.

4.6 The "Evidence" in Favor of the $H\phi O$ Effects in Protein Folding

In the 1960s and 1970s, all the available "evidence" for the $H\phi O$ solvation was the experimental data on the solubility of simple hydrocarbons in water — or, equivalently, the transfer Gibbs energies shown in Table 4.1. On the other hand, for pairwise hydrophobic groups, there were some data on the second virial coefficients and some data on the association of carboxylic acid. It was only in the early 1970s that the pairwise hydrophobic interaction was defined in terms of the PMF. This definition led to an experimental measure of the $H\phi O$ interaction.

During the years, much new "evidence" was accumulated. Perhaps the most complete list of "evidence" in favor of the $H\phi O$ effect is found in a review article by Dill *et al.* (2008):

> *There is considerable evidence that hydrophobic interactions must play a major role in protein folding: (a) Proteins have hydrophobic cores, implying nonpolar amino acids are driven to be sequestered from water. (b) Model compound studies show 1–2 kcal/mol for*

Table 4.1   Values of the Gibbs Energy Changes for the Transfer of Amino Acids from Cyclohexane to Water and from the Gaseous Phase to Water at 25°C [from Fersht (1999)]

Side chain of	Cyclohexane → H_2O (kcal/mol)	Vapor → H_2O (kcal/mol)
Leu	4.92	2.28
Ile	4.92	2.15
Val	4.04	1.99
Pro	3.58	1.50
Phe	2.98	−0.76
Met	2.35	−1.48
Trp	2.33	−5.88
Ala	1.81	1.94
Cys	1.28	−1.24
Gly	0.94	2.39
Tyr	−0.14	−6.11
Thr	−2.57	−4.88
Ser	−3.40	−5.06
His	−4.66	−10.27
Gln	−5.54	−9.38
Lys	−5.55	−9.52
Asn	−6.64	−9.68
Glu	−6.81	−10.24
Asp	−8.72	−10.95
Arg	−14.92	−19.92

transferring a hydrophobic side chain from water into oil-like media (234), and there are many of them. (c) Proteins are readily denatured in nonpolar solvents. (d) Sequences that are jumbled and retain only their correct hydrophobic and polar patterning fold to their expected native states (39, 98, 112, 118), in the absence of efforts to design packing, charges or hydrogen bonding.

To this list, one can add the "evidence" from experiments on the relative stability of mutated proteins. Let us examine each of these pieces of "evidence" separately (I hope that you noted the enclosing of the word "evidence" in quotation marks; you will see why below):

(a) The fact that the $H\phi O$ groups are in the interior of the protein was already noted by Kauzmann.

This fact in itself does not say that the $H\phi O$ interactions are *responsible* for bringing these groups to the interior of the protein. In general, one cannot conclude anything from *seeing* the result of an experiment. For instance, in the mixing process shown in Figure 4.5, we see mixing of two ideal gases. Most people rush to conclude that the positive change in entropy associated with this process is due to the *mixing process*. Such a conclusion is in general unwarranted. In the case of mixing, the conclusion is based on the erroneous association between positive entropy change and increase in the extent of disorder in the system. This topic is discussed in great detail in Ben-Naim (2008, 2012d).

In the case of protein folding, the mere fact that $H\phi O$ groups are found in the interior of the protein does not

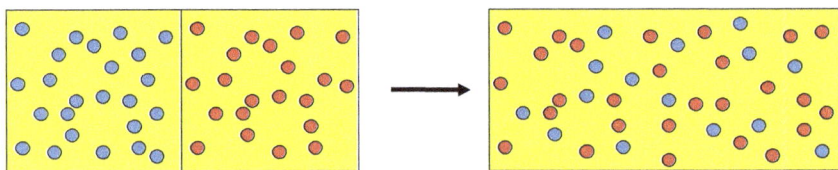

Fig. 4.5. Process of mixing of two different gases.

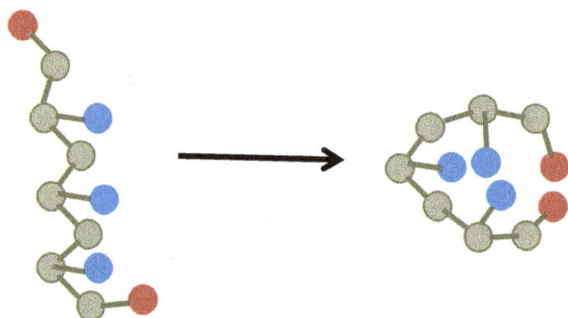

Fig. 4.6. Folding due to electrostatic attraction.

necessarily mean that the hydrophobic effects (either solvation or pairwise interaction) are responsible for the folding process.

Figure 4.6 shows schematically a segment of a protein with two charges of opposite signs on its two edges and a few $H\phi O$ side chains along the sequence. Clearly, because of the strong attraction between the two charges the protein will tend to fold in such a way that the two $H\phi O$ will be *forced* to occupy the interior of the folded structure. As we will see in the next section, the pairwise $H\phi I$ interactions are likely to be the most important factor which *forces* the $H\phi O$ groups to occupy the interior of the protein. Thus, from the fact that the $H\phi O$ groups are found in the interior of the protein, one cannot conclude that the $H\phi O$ interactions are responsible for bringing these groups to the interior of the protein. Therefore, "evidence" *a* in the above quotation is *no evidence*!

(b) Indeed, it is true that model compounds show that the transfer of an $H\phi O$ molecule from water into an organic liquid is associated with a negative change in Gibbs energy. Again, this argument was essentially Kauzmann's argument for the $H\phi O$ bond. Unfortunately, the Gibbs energy change for the transfer of a *molecule* such as methane, ethane or propane from water into an oil-like medium is *irrelevant* to the process of protein folding. First, the desolvation of a *molecule* is different from the desolvation of an $H\phi O$ *group* attached to the BB. In other words, the Gibbs energy of solvation should be replaced by the *conditional* Gibbs energy of solvation. The value of conditional Gibbs energies of solvation could be one or two orders of magnitude different from the solvation Gibbs energies of the same (or approximately the same) molecule.

For instance, the solvation Gibbs energy of methane in water is about 7.9 kJ/mol. The conditional solvation Gibbs energy of a methyl group attached to a saturated hydrocarbon BB is about 1.45 kJ/mol [Ben-Naim (2011d)].

Second, the solvation of an $H\phi O$ molecule in an oil-like medium does not even appear in the process of protein folding. The fact that the $H\phi O$ groups are found in the interior of the protein does not imply that a *solvation* Gibbs energy in the oil-like medium of the protein is involved in the process. As was shown almost 20 years ago, the transfer of an $H\phi O$ group attached to the protein BB involves *conditional desolvation* from water, and *interaction energy* between the $H\phi O$ group and its surroundings in the interior of the protein [see also Ben-Naim (2011d)]. The authors cite references to Wolfenden's article (2007), who in particular which is the "best" solvent to mimic the solvation in the oil-like media. Unfortunately, the solvation Gibbs energy of the solutes in the oil-like media is irrelevant to the protein folding.

Finally, "and there are many of them" in the quotation above refers to the many $H\phi O$ groups. As explained in Section 4.2, that was also part of Kauzmann's argument when he introduced the idea of the $H\phi O$ bond. Indeed, there are about 30% of side chain groups which are $H\phi O$. Each of these *does not* contribute 1–2 kcal/mol of Gibbs energy to the whole process of folding, but a much smaller quantity [incidentally, in item (b) of the quotation above, the authors say "show 1–2 kcal/mol for transferring…". They do not specify whether this is a contribution to the Gibbs energy change or the enthalpy change for the discussed process]. As we will see in Section 4.6, each of the $H\phi I$ effects is much larger that the corresponding $H\phi O$ effect, and *there are many more $H\phi I$ groups* than $H\phi O$ groups. As we noted earlier for a protein having 150 amino acids, there are about 50 $H\phi O$ groups. For the same protein there are about 350 $H\phi I$ groups.

For these reasons, "evidence" (a) and "evidence" (b) are not evidence "that hydrophobic interactions must play a major role in protein folding." It is regrettable that these two pieces of nonevidence appear in the literature almost 20 years after it was shown that they are in fact irrelevant to the process of protein folding, and that $H\phi I$ effects are far more important [see Ben-Naim (1989, 1990b, 2013) and Section 4.6].

(c) "Proteins are readily denatured in nonpolar solvents."

This fact is true for most nonpolar solvents, or even for addition of large quantities of a cosolvent such as urea or alcohol. But what has this fact got to do with evidence for the "major role in protein folding"? All it says is that the aqueous medium is *important* if not *essential* to the folding process.

It does not say anything about the relative importance of the $H\phi O$ effect to the overall driving force in the process of folding.

This evidence is likely a result of the erroneous assumption (or the belief in the myth) that water molecules affect the driving force for protein folding only through the $H\phi O$ effect. Once you replace the water by another solvent, or dilute the water in mixed solvents, you eliminate the $H\phi O$ effect, and therefore also eliminate the "major role" in the driving force for the folding. This assumption ignores other solvent-induced effects — specifically, $H\phi I$ effects, which are far more important than the corresponding $H\phi O$ effect. Thus, when we replace water by a "nonpolar solvent," the protein denatures not because of the elimination of the $H\phi O$ effects but, more importantly, because of the elimination of various $H\phi I$ effects (see Section 4.7). Therefore, we can conclude that "evidence" (c) in the quotation is anything but evidence!

(d) This is the most ridiculous "evidence" that "hydrophobic interactions must play a major role in protein folding." If you look at the references provided in the quotation, you will not find *any evidence* in favor of the $H\phi O$ effect in protein folding. The references are: Cordes *et al.* (1996), Hecht *et al.* (2004), Kamtekar *et al.* (1993), Kim *et al.* (1998). Unfortunately, none of these references provide any support for the claim made by Dill *et al.* (2008) in the quotation.

Cordes *et al.* (1996), in an article titled "Sequence Space, Folding and Protein Design," asked the following question: "Which features of a protein's sequence are most important in determining the structure?" They considered several answers to this question, eventually reaching the following conclusion:

> *The most important design elements seem to be the proper element of hydrophobic residues along the polypeptide chain and the ability of these residues to form a well packed core. Buried polar interactions turn capping motifs and secondary structural propensities also contribute, although probably to a lesser extent.*

I thought the findings of those articles very interesting. However, I do not believe that by studying mutated proteins one can reach any meaningful conclusion regarding the relative importance of the different types of interactions. Each mutation can cause changes in the many types of interactions; see below. Therefore, the citation of this article by Dill *et al.* (2008) as "providing" considerable evidence that hydrophobic interactions must play a major role in protein folding is misleading. In fact, this "evidence" does not even say anything about the role of water in protein folding, and therefore it was deemed to be "not even wrong" [Ben-Naim (2011d, 2013)]. Thus, we see that all the "considerable evidence" listed by Dill *et al.* is in fact nonevidence.

There is one more evidence which is discussed in the literature and involves mutations of proteins. A summary of this argument may be found in a recent book by Kessel and Ben-Tal (2011).

> *The dominance of non-polar interactions as a driving force of protein folding has been demonstrated by a wide range of studies, from simple observations of protein denaturation by organic solvent, to complex thermodynamic and spectroscopic measurements. This dominance is so profound, that the overall structure of a protein can be maintained even after its sequence is randomized, as long*

as the original hydrophobic–hydrophilic pattern is kept. While virtually all structural-biophysicists agree that the hydrophobic effect (i.e., non-polar interactions) is the major driving force of protein folding, they disagree on the magnitude of this effect. One way to investigate this is to mutate specific residues participating in non-polar interactions, and to see how the free energy of denaturation changes compared to that of the original protein.

In the first part of this quotation, the authors reiterate the standard dogma about the dominance of the hydrophobic effect. In the last part, the authors allude to another evidence based on experimental data on the relative stability of mutated proteins.

The idea is to synthesize modified (or mutated) proteins where one or two amino acids are replaced by other amino acids. The simplest example is replacing one $H\phi O$ group by one $H\phi I$ group, or vice versa, and then measuring the Gibbs energy change for the folding process. For concreteness, suppose that we replace one $H\phi I$ group which is known to be in the interior of the protein by an $H\phi O$ group. We find that the standard Gibbs energy of the process of folding becomes more negative, i.e. the folded form of the mutated protein is more stable relative to the unfolded form than the original protein. Can one say anything about the role of the $H\phi O$ effect on the stability of the protein?

In my opinion, it is extremely difficult to conclude anything regarding the relative importance of $H\phi O$ and $H\phi I$ groups from such measurements. Most biochemists maintain otherwise. For instance, Pace *et al.* (2014) write:

To study the contribution of the hydrophobic interactions to protein stability, a variant is made in which a buried hydrophobic group *is removed and then the stabilities of the wild type and the variant protein are measured to find the change in the stability which is denoted a $\Delta(\Delta G)$ value.*

Their conclusion:

Hydrophobic interactions and hydrogen bonds both make large contributions to protein stability.

I fully agree with this conclusion. However, I have two reservations about it. First, I believe that HBs as well as other possible $H\phi I$ effects contribute more than $H\phi O$ effects to the stability of proteins. Second, I disagree with the methodology used in this article (as well as many others) to study the relative contribution of the $H\phi O$ and $H\phi I$ effects.

In my view, it is almost impossible to conclude anything from such experiments on the role of the $H\phi O$ effect on the stability of protein. The reasons are plentiful.

First, any change of even a single amino acid can cause a large change in the 3D structure of the entire protein. In such a case, all the $H\phi O$ and $H\phi I$ effects due to all the groups of the protein might change, and it will be impossible to isolate the effect of the specific amino acid which was replaced.

Second, suppose we know that the overall structure of the protein does not change significantly, but some small local changes can occur in the vicinity of the replaced amino acid. This small local change can affect some of the strength of the $H\phi I$ interactions, either weakening or strengthening the interaction — causing either an increase or a decrease in the standard Gibbs energy of the folding process. Note also that replacing one amino acid by another would affect the solvation of both the folded form and the unfolded form, and these effects could be in the opposite direction.

Therefore, in any analysis of the net effect, one must take into account the effect of the new amino acid on the stability (i.e. the solvation Gibbs energy) of both the folded and the unfolded form.

Finally, suppose that the replacement of, say, an $H\phi O$ group by an $H\phi I$ group of the same size did not have any effect on the structure of the protein. We find that ΔG of the folding has increased. Does this mean that the $H\phi O$ effect which was eliminated is important? The answer to this question cannot be given without knowing exactly how the replacement of the $H\phi O$ group has changed *all the interactions* between the other groups in the vicinity of the replaced group.

Figure 4.7 shows a schematic example where a replacement of a CH_3 by OH can cause the formation of a new HB (hence increasing the stability of the protein) or disrupt an existent HB between two $H\phi I$ groups (hence decreasing the stability of the protein).

Thus, in general, one cannot conclude much from such experiments on mutated proteins.

Most of what we said above referred to $H\phi O$ solvation of molecules or groups attached to the protein. One can similarly discuss the relevance of the PMF between two $H\phi O$ molecules to the stability of the protein. Again, much work has been done on the PMF between pairs of $H\phi O$ molecules — say, two methane molecules in water. These studies are of interest in the context of the study of the properties of aqueous solutions, but they are totally irrelevant

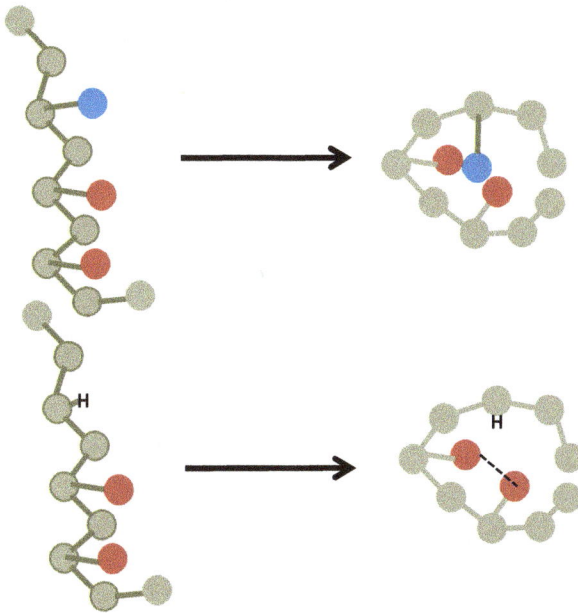

Fig. 4.7(a) Example of mutation.

to the study of protein folding. As has been shown in the late 1980s, the relevant quantities are the PMF of two $H\phi O$ groups attached to a BB. This has been referred to as the *conditional* PMF between two $H\phi O$ groups.

It is regrettable that arguments based on the PMF between two $H\phi O$ groups still appear in the literature almost 20 years after the strong evidence in favor of the $H\phi I$ effects was published! As an example, in a recent article by Dias and Chan (2014), we find:

Simulations of pairwise hydrophobic interactions have provided rationalizations for several intriguing phenomena in protein folding. ... pairwise hydrophobic interactions between solutes of different sizes can yield critical insight into folding energetics.

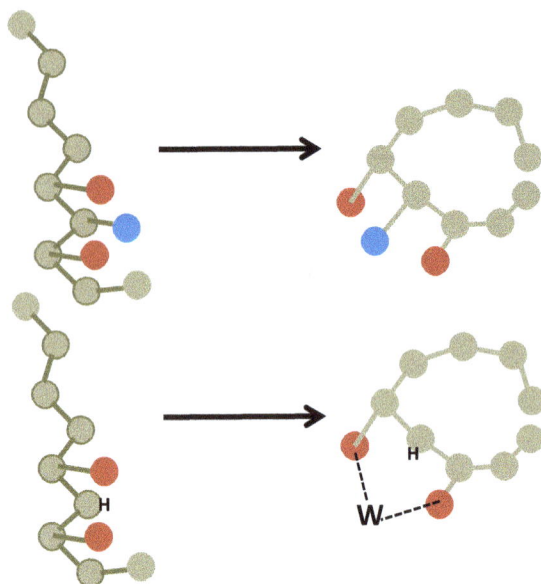

Fig. 4.7(b) Example of mutation alanine (blue) being replaced by glycine.

Such statements echoing the traditional dogma completely ignore the *fact* that hydrophobic pair interactions are by themselves *irrelevant* to protein folding. These interactions do not provide "rationalizations" or "insights" into protein folding. The statements also ignore the fact that pair interactions between *hydrophilic* groups are far more important to the process of protein folding.

It should be emphasized again that the solvation of $H\phi O$ molecules, as well as the PMF between two $H\phi O$ groups, should be studied in the context of the study of the properties of aqueous solutions [Ben-Naim (2009, 2011d)]. These studies are totally irrelevant to protein folding or protein–protein association unless one takes into account the effect of the BB of the protein.

4.7 The Evidence in Favor of the $H\phi I$ Effects in Protein Folding

In this section, we summarize briefly the evidence in favor of the $H\phi I$ role in the stability of the protein. There are essentially three $H\phi I$ effects on which we have enough data to support the claim that $H\phi I$ effects are far more important than $H\phi O$ effects in protein folding.

First, the transfer of an $H\phi I$ group from being exposed to water in the unfolded form to the interior of the protein in the folded form (Figure 4.3). This $H\phi I$ effect is the analog of the Kauzmann model for transferring an $H\phi O$ group into the interior of the protein.

The desolvation of a hydrophilic group attached to the protein in the U form depends on the number of "arms" along which it can form HBs with water molecules. An NH group has one arm and the conditional solvation Gibbs energy of such a group is about -2.25 kcal/mol. A carbonyl group has two "arms" and its conditional solvation Gibbs energy is about -4.5 kcal/mol. A hydroxyl group has three arms and the corresponding conditional Gibbs energy is about -6.75 kcal/mol.

It is known that most of the $H\phi I$ groups which can form HBs are indeed hydrogen-bonded when they are in the interior of the protein. Thus, for the formation of one genuine HB by two arms that were desolvated (Figure 4.8), we have a net change in Gibbs energy of about

$$\delta G_2^{HB} = \varepsilon_{HB} - 2\Delta G^*_{one\ arm} \approx -6.5 - 2 \times (-2.25)$$
$$\approx -2 \text{ kcal/mol}, \tag{4.3}$$

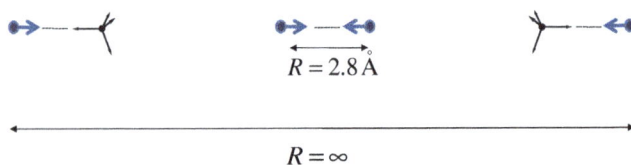

Fig. 4.8. The process of formation of a direct HB between two hydrophilic groups.

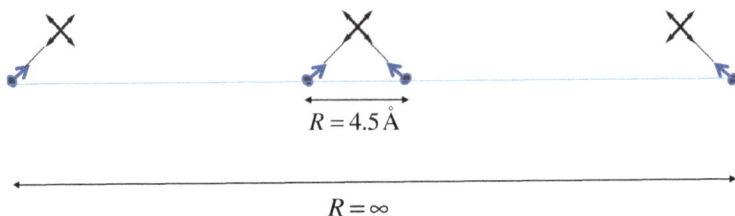

Fig. 4.9. The process of formation of an indirect HB bridge between two hydrophilic groups.

where ε_{HB} is the HB energy, and $\Delta G^*_{\text{one arm}}$ is the conditional solvation Gibbs energy per arm (note that, unlike the conditional solvation Gibbs energy of an $H\phi O$ group, which strongly depends on the type of BB, the corresponding quantity for $H\phi I$ is less sensitive to the type of BB).

The second $H\phi I$ effect is the conditional pairwise PMF between two $H\phi I$ groups at a specific distance and orientation as shown in Figure 4.9. This is a relatively new effect. Its origin is explained in Ben-Naim (2009, 2011d). The net contribution of this effect to the stability of the protein is about

$$\delta G_2^{H\phi I} = \Delta G^*_{1,2}(R = 4.5\text{Å}) - 2\Delta G^*_{\text{one arm}}$$

$$\approx -2.5 \, \text{kcal/mol}. \tag{4.4}$$

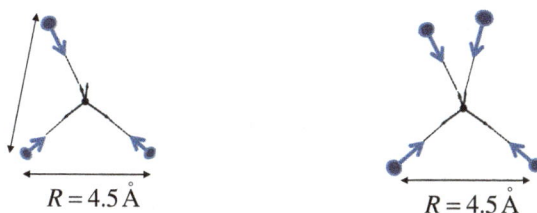

Fig. 4.10. Hydrophilic interaction between three- and four-hydrophilic groups.

Here, $\delta G_2^{H\phi I}$ is the solvent-induced part of the Gibbs energy change for bringing the two $H\phi I$ groups from infinite separation to a distance of about 4.5 Å and with the orientation of the two arms such that a water molecule can form an HB bridge between the two arms. $\Delta G_{1,2}^*(R = 4.5\text{Å})$ is the solvation Gibbs energy of pairs of $H\phi I$ groups at the configuration shown in Figure 4.9.

This $H\phi I$ effect was discovered almost 25 years ago. The estimated value of $\delta G_2^{H\phi I}$ is far larger than the corresponding pairwise conditional PMF between two $H\phi O$ groups.

Similarly, one can have $H\phi I$ interactions between three $H\phi I$ groups, or even four. These are shown in Figure 4.10 and are discussed in great detail elsewhere [Ben-Naim (2010)].

Regarding the evidence for the magnitude of the various $H\phi I$ effects, these are discussed in Ben-Naim (2011d). Here, I will only list the sources of evidence:

1. The second peak in the radial distribution function of water at $R = 4.5$ Å.

2. Theoretical work based on a model of water molecules described in Appendix A. The first estimates were made in 1989 and repeated in detail in Ben-Naim (2009).
3. Experimental data. (For details, see Ben-Naim (2011d)).
4. Simulated experiments on two water molecules at a fixed orientation approaching each other within water.

All this evidence supports the original theoretical estimates of the strength of the $H\phi I$ interactions.

I should add here that very recently Busch *et al.* (2013), in an article titled "Water Mediation Essential to Nucleation of β-Turn Formation in Peptide Folding Motifs," concluded:

> For GPG-NH$_2$ in water, hydrogen-bonding interactions appear to be the primary driving force in inducing this common β-turn sequence to fold. It is highly likely that hydrophilic forces are just as important in driving protein folding as the hydrophobic effect in solution, especially for the initiation of this process in vivo.

We can conclude that each $H\phi I$ effect is much stronger than the corresponding $H\phi O$ effect. In addition, the total number of $H\phi I$ groups is far larger than the total number of $H\phi O$ groups. Therefore, one can safely conclude that the various $H\phi I$ effects play a dominant role in determining the stability of the structure of proteins. [For details, see Ben-Naim (2011d, 2013).]

In the next section, we will briefly review some applications of $H\phi I$ effects in key biochemical processes involving proteins.

4.8 The Relevance of the $H\phi I$ Effects to Biochemical Processes Involving Proteins

We have seen that each $H\phi I$ effect is larger than the corresponding $H\phi O$ effect. One can argue that the total $H\phi O$ effects in a real protein could be larger than the total $H\phi I$ effects. This could be true in principle, but it is very unlikely. First, we do not know how to estimate the total $H\phi O$ effects or the total $H\phi I$ effects. These solvent-induced effects are not additive [see Ben-Naim (2011d)]. But, even if we assume additivity, there are far more $H\phi I$ groups than $H\phi O$ groups in protein. For instance, in a protein of 150 amino acids there are about 50 $H\phi O$ side chains, but over 300 $H\phi I$ groups.

The various contributions of the $H\phi I$ effects (both δG_2^{HB} and $\delta G_2^{H\phi I}$; see Section 4.6) are explained in great detail in Ben-Naim (2013). Here, we add that $H\phi I$ effects can also contribute to our understanding of the pressure effect on protein denaturation. Another phenomenon of great interest (and perhaps some mystery as well) is the phenomenon of cold denaturation.

This is traditionally explained by invoking the almighty $H\phi O$ effect. However, it was recently shown that the difference in the temperature dependence of the two $H\phi I$ effects, δG_2^{HB} and $\delta G_2^{H\phi I}$, could account for both heat and cold denaturation [Ben-Naim (2013)].

A second, not less important process in which $H\phi I$ effects can contribute significantly is the self-assembly of proteins.

In an editorial of *Science* in 2005, one finds another "big question" of science:

"How do proteins find their partners?" Protein–protein interactions are at the heart of life. To understand how partners come together in precise orientations in seconds, researchers need to know more about the cell's biochemistry and structural organization.

Again, this question could not be answered within the $H\phi O$ paradigm. It was therefore deemed to be one of the "big questions" of science. Indeed, the process of protein–protein association to form a stable and long-lived dimer (or higher aggregate) is not less of a mystery than the process of protein folding.

In forming a stable dimer, one monomer loses its translational entropy. This is true of any association process. In most cases of dimerization, a chemical bond is formed between the monomers. The energy of the bonds compensates for the loss of translational entropy, making the dimerization process favorable. In protein association, there is no covalent bond between the monomers. Hence the mystery: What is the factor that compensates for the loss of translational entropy? In some textbooks, you might find the answer that the $H\phi O$ effect is the dominant factor in this process. However, a careful examination of the process shows that this factor is minor and cannot account for the stability of the dimer (or of any other aggregate, for that matter). On the other hand, the inclusion of the $H\phi I$ effect shows how an extremely unlikely process of dimerization turns into an extremely likely process. The details of the $H\phi I$ contribution to the process of association may be found in Ben-Naim (2011d). Here, we provide only a qualitative description of the basic idea.

Consider two globular proteins in contact. Clearly, the direct interaction between the two monomers is minor. The Gibbs energy for this process will be large and positive. Next, we take into account the solvent-induced effect. We start with completely $H\phi O$ monomers, i.e. proteins whose surface consists of the $H\phi O$ group. We find that in this case the Gibbs energy of the process of dimerization is large and positive. This means an unfavorable dimerization process.

Next, we gradually increase the mole fraction of $H\phi I$ groups on the surface of the protein. Figure 4.11 shows the

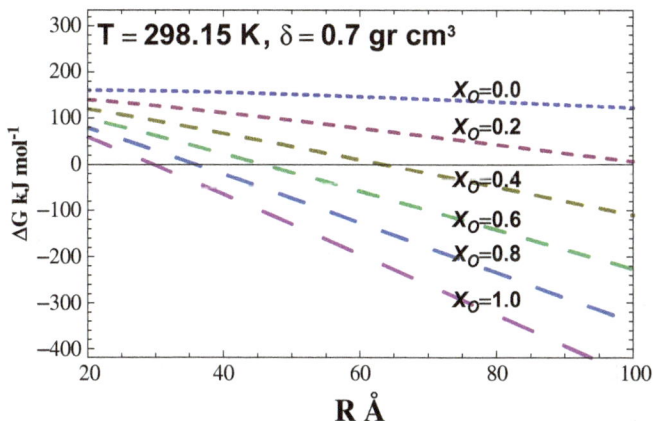

Fig. 4.11. Dependence of the Gibbs energy of association for different values of the mole fraction of hydrophilic groups on the surface of a protein.

dependence of the Gibbs energy of dimerization as a function of the radius of the protein for several mole fractions of the $H\phi I$ groups. We see that once the mole fraction of the $H\phi I$ group is larger than about 0.4 or 0.6, the curve turns into a negative territory, i.e. making the process of dimerization favorable. The larger the mole fraction x_0, the more negative the Gibbs energy change.

On the molecular level, the reason for the large negative Gibbs energy of dimerization is simple to visualize. When the two proteins are in contact, there exists a strip on the surface of each monomer from which water bridges between $H\phi I$ groups may be formed (Figure 4.12). It was estimated that each water bridge between two $H\phi I$

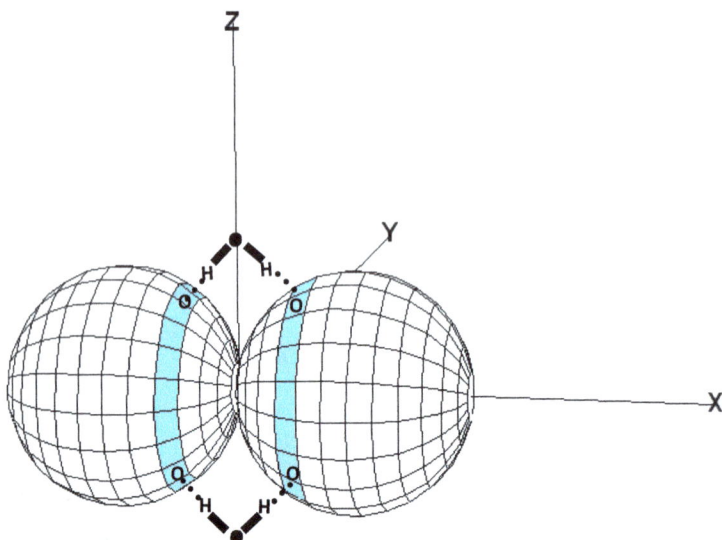

Fig. 4.12. Two water bridges between two globular proteins.

groups can contribute about -11 kJ/mol. This effect is almost an order of magnitude larger than the corresponding $H\phi O$ effect. Clearly, as the proteins become larger the number of $H\phi I$ groups that can be bridged by water molecules becomes larger too, as can be seen from Figure 4.11. For proteins with a radius of about 50 Å, the Gibbs energy of dimerization becomes large and negative. This finding solves the mystery associated with the question of the stability of an assembly of proteins.

So far, we have shown how $H\phi I$ effects (mainly δG_2^{HB} and $\delta G_2^{H\phi I}$) contribute significantly to the stability of the dimers or higher aggregates of protein. A not-less-important question is the specificity of the binding mode of the two or more proteins. This is known as molecular recognition. The $H\phi I$ effects contribute not only to the Gibbs energy of the aggregation process but also to the selection of the specific binding mode. The pairwise $H\phi I$ effects can also contribute to the molecular recognition in the processes of binding drugs to proteins and to DNA. This is a new and potentially revolutionary approach to drug design. [Further details are provided in Ben-Naim (2011d).]

Finally, the reader should note another well-known phenomenon, which is the *solubility* of globular proteins.

The high solubility of proteins is another mystery that did not get sufficient attention in the literature. The reason why the solubility of protein was not included in the list of "big questions" of science is simple. Everyone who has given a thought to the question regarding the reasons for the large solubility of globular proteins in water correctly concluded that the hydrophilic groups on the surface of the proteins are responsible for their solubility. This is of course true — but it is not the whole truth! What was not known is that it is not the mere *number* of hydrophilic groups exposed to the solvent that determines the high solubility of the proteins, but the *specific distribution* of the hydrophilic groups on the surface of the protein. It was recently shown that the possibility of formation of water bridges between two or more hydrophilic groups contributes significantly, if not decisively, to the large solubility of the protein in water [Ben-Naim (2011d)].

4.9 Conclusion

For over 50 years, the $H\phi O$ paradigm reigned supreme in the biochemical literature. It thrived not on evidence but on faith and on lack of alternatives. This situation changed radically when a complete inventory of all solvent-induced effects was obtained [Ben-Naim (1989, 1990a, 1990b)]. This led to two important conclusions: first, the conventional $H\phi O$ solvation and $H\phi O$ interaction were found to be totally irrelevant to the process of protein folding; second, a host of new $H\phi I$ effects were discovered and shown to be much stronger than the corresponding $H\phi O$ effects. The resulting change of paradigm from $H\phi O$ to $H\phi I$ has not only demystified many biochemical processes, but in fact provided answers to long-standing problems, such as the process of protein folding, self-assembly of proteins and molecular recognition.

Today, it is clear that there is overwhelming evidence for the importance of the $H\phi I$ effects — not only in providing stability to the structure and the assembly of proteins, but also in providing strong *forces*, which explains the mystery of the fast-folding protein. We will discuss these forces in the next chapter.

5

From Levinthal's Question to Resolving Levinthal's "Paradox"

5.1 Abstract

In early 1973, Levinthal asked a simple question: What are the factors that *speed* and *guide* the proteins in the process of protein folding? This question became one of the "big questions" of science. Levinthal tried to answer it. Assuming that a protein walks randomly in its configurational space, it would take eons to reach the native structure. Therefore, Levinthal concluded that there must be some "local interactions" that speeded and guided the protein toward the native structure. In his writings, there is no hint of any paradox. Nevertheless, for many years Levinthal's *question* and his tentative answer were considered to be a paradox. Many scientists tried to resolve the paradox, which actually never existed. This consumed a great deal of effort. The apparent paradox is that proteins fold in a very short time — not in eons, as estimated from a random walk. However, there is no paradox and there was never a paradox. The immediate answer to Levinthal's question is simple: the rapid and guided folding of the protein must result from some *strong forces* that are exerted on the groups of the protein. The main question is, therefore, to find out what the origin of these strong forces is. Unfortunately, people stuck to the paradigm of the "dominant $H\phi O$ forces," which could not provide any strong *forces* resulting from the $H\phi O$ interactions. This chapter discusses the origin of the *strong forces* that speed and guide the folding of the protein. The missing factor in Levinthal's question is now available — the strong solvent-induced force operating on $H\phi I$ groups. The discovery of the $H\phi I$ forces essentially resolves the dynamic aspect in the PFP.

5.2 The Origin of the Myth and Some Historical Notes

As we have noted earlier, there are essentially two problems associated with the process of protein folding. The first is concerned with the factors that impart stability to the native structure of the protein. The second is concerned with the questions of how and why protein folds to its native 3D structure in a very short time.

These two questions presented formidable challenges to chemists, biochemists and physicists. In this chapter, we focus only on the second question — the one referred to in the quotation from the *Science* editorial (2005). Let us start with a quotation from Levinthal (1968, 1969):

> *Let us ask ourselves how proteins fold to give such a unique structure.*
> *By going to a state of lowest free energy? Most people would say yes*

and indeed, this is a very logical assumption. On the other hand, let us consider the possibility that it isn't so.

(a) *How accurately must we know the bond angles to be able to estimate these energies? Even if we knew these angles to better than a tenth of a radian, there would be 10^{300} possible configurations in our theoretical protein. In nature, proteins apparently do not sample all of these possible configurations since they fold in a few seconds, and even postulating a minimum time for going from one conformation to another, the proteins would have time to try the order of 10^8 different conformations at most before reaching the final state.*

(b) *We feel that protein folding is speeded and guided by the rapid formation of local interactions which then determine the further folding of the peptide. This suggests local amino acid sequences which form stable interactions and serve as nucleation points in the folding process.*

(c) *Then, is the final configuration necessarily the one of the lowest free energy? We do not feel that it has to be. It obviously must be a metastable state which is in a sufficiently deep energy well to survive possible perturbations in a biological system. If it is the lowest energy state, we feel it must be the result of biological evolution, i.e. the first deep metastable trough reached during evolution happened to be the lowest energy state. You may then ask the question, "Is it a unique folding necessary for any random 150-amino acid sequence?" and I would answer, "Probably not."*

The propositions above are cast in a form which is reminiscent of a mathematical proof by contradiction; let us assume that X is true, reach an absurd result, then conclude that our assumption cannot be true. In fact, we have here all the elements of such a proof.

Statement (a) raises two questions: (1) How does a protein fold to give a unique structure? (2) Is this unique structure the state of the lowest free energy? We will refer to the first question as the *Levinthal question*. The second question, as well as its answer, will not be discussed here. It is related to Anfinsen's thermodynamic hypothesis, and it is discussed in Chapter 6.

Statement (b) essentially concludes that if one assumes a random sampling of the configuration space of the protein, then one arrives at an absurd result. Statements (c) and (d) suggest possible answers to the two questions raised in statement (a).

Clearly, there exists no *paradox* in obtaining an absurd result based on an unrealistic assumption. A paradox is defined as an unacceptable conclusion derived by apparently acceptable reasoning from apparently acceptable premises. Most people who wrote about Levinthal's paradox interpreted Levinthal's writings as if it implied that the protein walked randomly in its configurational space. In fact, Levinthal said that "proteins apparently do not sample all of these possible configurations...." It is clear that he did not assume a random search in the configurational space. Certainly, he did not envisage a flat, golf-course like energy landscape, as is commonly described in the literature [e.g. Dill *et al.* (2008)].

Instead, Levinthal immediately recognized that the absurd result he reached followed from the *wrong assumption* of a random search over the immense configurational space. He did not see that absurd result as a paradox, as so many others did. He immediately reached the (almost) correct solution, as given in statement (c). Namely, that there must be *preferential pathways* of folding, "guided by rapid formation of local interactions." Although Levinthal did not specify what these "guiding interactions" are, his solution to the absurd result (based on unrealistic assumption) is almost correct. Instead of "guiding interactions," one should use the term "guiding forces." Though these *forces* are derived from the *interactions*, it is the magnitude of the *force* acting on the groups of the protein that determines the speed of the folding process. The main question left unanswered by Levinthal is this: What are these strong forces that guide the protein to fold into its native structure in a relatively short time? We can now answer Levinthal's question by claiming that these forces originate from the water — more specifically, the solvent-induced forces exerted on the hydrophilic groups along the BB and the side chains of the protein. [For details, see Ben-Naim (2009, 2011d).]

Levinthal was probably the first scientist to express doubts about what became another dogma. In statement (d), he asked, "Is the final configuration necessarily the one of the lowest free energy?" Then he proposed the answer: "We do not feel that it has to be." Levinthal raised doubts regarding the "lowest free energy," but he did not provide any arguments as to why the native structure should not necessarily be at the lowest free energy. We will further discuss this question and its answer in Chapter 6.

5.3 Attempts to Resolve the Levinthal Paradox

As noted in Section 5.1, for over 40 years people endeavored to "resolve" the Levinthal paradox instead of trying to find an *answer* to Levinthal's question.

Perhaps the most serious and much-acclaimed attempt to "resolve" the paradox was published by Zwanzig *et al.* (1992). In their introduction, Zwanzig *et al.* write:

> *The main point of this paper is to show by mathematical analysis of a simple model that Levinthal's paradox becomes irrelevant to protein folding when some of the interactions between amino acids are taken into account.*

This is exactly the answer given by Levinthal himself; namely, that the *interactions* between different parts of the protein can *guide* the folding process. As we have pointed out above, the important guiding factors are the *forces* rather than the interactions. Zwanzig *et al.* did not offer any answer to the question regarding these forces, nor did they specify which "some of the interactions between amino acids" are. Furthermore, the model used by them is not a realistic one, and might even be misleading.

It is strange that while Zwanzig *et al.* recognized that "Levinthal concluded that the random searches are not an effective way of finding the correct state of folded protein," which essentially means that Levinthal did not consider the assumption of a random search to be correct. These authors concluded, "Nevertheless, proteins do fold, and in a timescale of seconds or less. This is the paradox."

In my view, there was never a paradox, and Levinthal never considered that as a paradox.

Zwanzig *et al.* drew on Dawkins' brilliant ideas of explaining the mechanism of evolution. Briefly, the protein is viewed as a sequence of N bonds, and the "connecting bond between two neighboring amino acids can be characterized as 'correct' or 'incorrect.'" ("Correct" means "native" in biology.) Then they assumed some rate constant (k_0) for the transition "correct"→"incorrect," and another rate constant (k_1) for the transition "incorrect"→"correct." Assuming further that the ratio (k_0/k_1) is small, they calculated the mean first-passage time to reach the fully "correct" information.

It should be noted that the metaphor used by Dawkins is barely suitable for explaining evolution to the layperson. The mechanism arriving at the "correct" target as proposed by Dawkins demonstrates the *possibility* of occurrence of an event which is perceived to be highly improbable. As such, Dawkins' model achieves its goal of removing the mystery from the evolutionary process. However, even in evolution, there exist no "correct" or "incorrect" results. In fact, Dawkins himself recognized that his explanation is not relevant to the actual process of evolution. Evolution does not set any *goals* or *targets* to reach. Nevertheless, one can simply *define* a "correct" outcome as one which has some evolutionary advantage. This is not the case for the protein folding process. Therefore, Dawkins' metaphor is not adequate for the process of protein folding. The main objection to this model is that one cannot justify the preferential transition from "an incorrect bond" to "a correct bond" at each stage of the protein folding process.

In evolution, the metaphor of transitions from "incorrect" to "correct" is biased according to some selection criterion, i.e. the "correct" result has some advantage, and therefore that result survives. There exists no analog of the selection criterion in the process of protein folding. [More on this in Ben-Naim (2013).]

When a protein folds, it does not "know" which is the "correct" way to go, and besides there is no way to define the "correct bond" for each bond in a protein.

Furthermore, Zwanzig *et al.* did not provide a plausible reason for the particular assignment of the values of the rate constants k_0 and k_1 in terms of either molecular interactions or forces. Therefore, the model used by them, as well as the specific solution of the model, is not relevant to the PFP.

There are other statements involving evolution theory and the PFP. In a recent article, Wolynes (2005) writes:

> *Evolution solved the protein-folding problem. A major goal of biomolecular science has been to understand how this was done.*

Of course, evolution does not *solve* any problem, nor was the PFP posed to Nature. Evolution only *evolves* and a product which has some evolutionary advantage survives. It is clear that Wolynes means "solved" in his first sentence only in a figurative sense. In the same sense, people say today that some bacteria "developed" resistance to some drug. Of course, bacteria do not "develop" anything. In a given population there might be many mutants of the same bacteria, and some might be resistant to a specific drug. When that drug is administered, only those mutants that are resistant to the drug will survive. To an outside observer, it looks as if the population as a whole had "developed" the resistance to the drug.

Thus, it is acceptable to use the word "develop" in the sense that this is how it seems to an observer who is not aware of the existence of resistant mutants in the original population. However, it is meaningless to try to "learn" from the bacteria (or from evolution) how they developed the resistance to the drug.

Similarly, during evolution, proteins — or, rather, polypeptides — were synthesized. Some folded, while others did not. Of those that folded, some reached a stable 3D structure, and some did not. Of those that reached a stable structure, some had some advantage, while others did not, and so on.

Looking at the final outcome of a functional protein, one can say figuratively that Nature or evolution has "solved" the problem of folding a polypeptide into some useful 3D structure. This is acceptable only if we understand that in the "population" of all the peptides, some have folded into a useful 3D structure. Not because this structure was the target of evolution, and not because evolution had faced the problem of how to fold a specific protein. However, one cannot attempt to *understand* how evolution has "solved" the PFP, simply because evolution did not solve any problem.

There are many other speculations involving the role of evolution in determining the shape of the Gibbs energy landscape (GEL), the speed of folding, etc. I do not believe that these speculations are useful for understanding the PFP.

Returning to the folding of proteins, we can say that, as in evolution, proteins do not have a *target* to reach. Instead, they respond to forces exerted on the groups of protein, which both guide and speed the folding of the protein.

5.4 Target-Based Versus Cause-Based Processes

Suppose that a drunken man walks randomly in the street of a very big city. We ask: What is the probability that he will reach a *specific* building, say point X, starting from point A, in an hour? Clearly, if the city is very big, the probability of reaching a *specific* point X within an hour is extremely small — the same as the probability of a monkey typing a *specific* sentence from *Hamlet*. This situation is schematically depicted in Figure 5.1.

Next, we ask: What is the probability of reaching *any* building in the city, given that the drunken person started from the same point A, and wandered randomly? In this case, the probability that he will reach *any* building, say Y, within an hour, is nearly 1 (Figure 5.2).

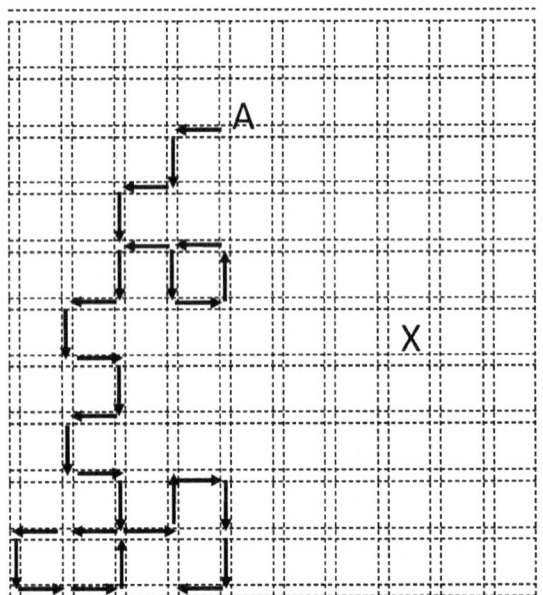

Fig. 5.1. A random walk in a big city, starting at A.

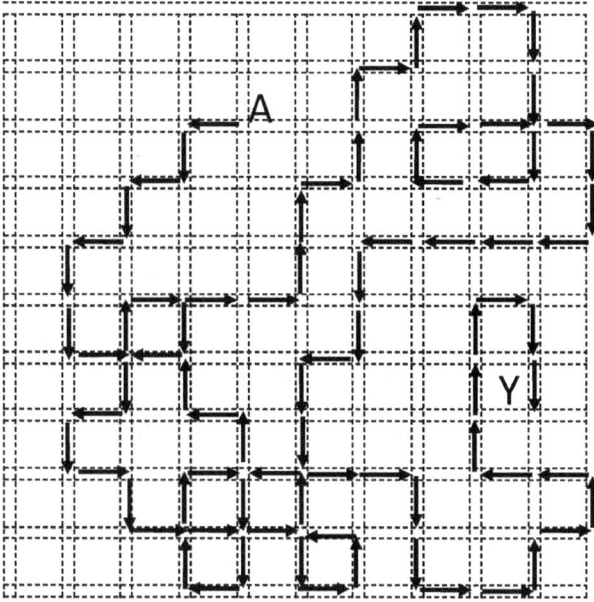

Fig. 5.2. A random walk in a big city.

Next, we repeat the experiment and ask: What is the probability of reaching again the point Y starting from A? The answer is the same as in the first experiment with the *specific* point X.

Now, the experiment is repeated several times and we find that the drunken man when released from several different initial points A will reach the same point Y. This seems like a paradox if we assume that he walks randomly. Levinthal was also puzzled by the fact that a protein starting from the denatured state (A) reaches the same "target" point of the native 3D structure (Y), although the probability of reaching Y from A by a random walk is extremely low. He did not consider his absurd result to be a paradox, or seek a solution to the "paradox," as many others did. He correctly *abandoned* the *assumption* that the protein walks at random in the configurational space. Levinthal's conclusion that the protein does not "walk" totally at random is essentially correct. What he left unanswered is this question: *What* are the factors that speed and guide the protein to fold in a narrow range of pathways?

Going back to the metaphor of the walking drunken man, if we find that always, or most of the time, he reaches Y from the initial point A, we might believe that he walks toward some "target." However, if we search for a scientific explanation of this phenomenon, then we should either question the validity of the random walk assumption or assume that there exist some other factors that "speed" and "guide" the man along some preferential pathway.

In the case of the drunken man, there is clearly no "target" toward which he walks! However, we can imagine a pattern of backwinds that *pushes* him along some preferred directions. This pattern of directions does not have to be always pointing toward the target. At some streets the wind can force the person to veer away from the target, but the net effect of this pattern of winds will bring him to the point Y. We will refer to this factor as the "cause-directed" one. (See Figure 5.3.)

In the case of protein folding, Levinthal correctly concluded that there must be some factor that "speeded" and "guided" the protein. He did not specify whether the factor is target-guided or cause-guided. It is unfortunate that the great majority of researchers searched in the wrong direction. First, because of misinterpretation of Anfinsen's thermodynamic hypothesis (see Chapter 6), people searched for a global minimum in the Gibbs energy landscape toward which

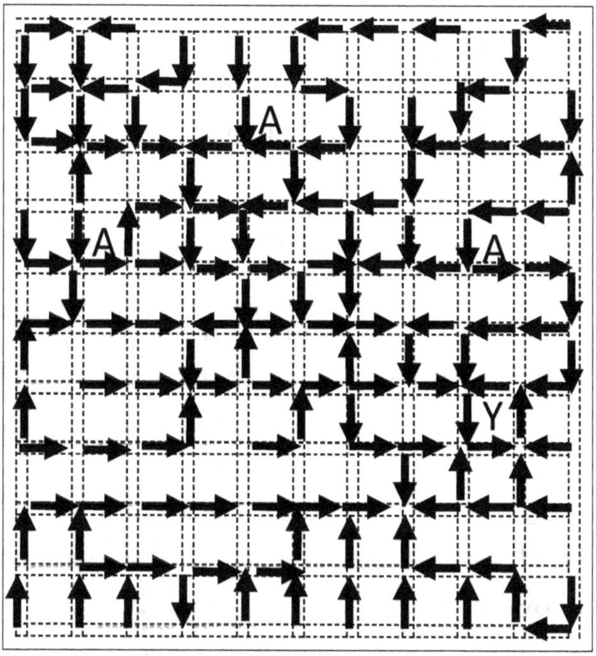

Fig. 5.3. A guided walk from any point A toward Y.

the protein is attracted. Second, from Anfinsen's hypothesis people speculated about the existence of some code translating from the sequence into the native structure, and all we have to do is to decipher this code, and then we will understand how the protein "knows" which target it must reach.

5.5 What Is the Answer to Levinthal's Question?

Suppose that we make the following experiment. We have two metal balls held at two points at a distance R from each other (Figure 5.4). We release one of the balls, numbered 2, and we observe that it flies *directly* and at high speed toward the second ball, numbered 1. If we repeat the experiment

Fig. 5.4. Two balls in a vacuum.

again and again, we find that the first ball always flies directly toward the second ball. Why?

A target-based theory would claim that the ball is biased to fly in the "correct" direction. Or that evolution has "solved" the problem of how to guide and speed the ball toward its target.

On the other hand, a cause-based theory would look for the *force* which both directs and speeds the flying of the ball. Once we know the forces, we can repeat the experiment with different spheres affected by different forces. We will find that the larger the force the greater the speed of the flying.

When the same experiment is carried out in a solvent, the situation is more complicated. Here, in addition to the *direct* force exerted by one ball on the other, there are many more molecules in the solvent that also exert forces on the balls.

Now, suppose that we do the same experiment in vacuum or in some organic liquid, and we find that the released ball does not necessarily fly directly toward the second ball. In a solvent, we might see some random motion (Figure 5.5) which is typical of the motion of Brownian particles. Once in a while the ball might reach the other ball, but this is a very rare event. We can conclude that there is no strong direct force acting between these two balls.

Finally, we repeat the same experiment with the same two balls as in the previous experiment, but now we choose the

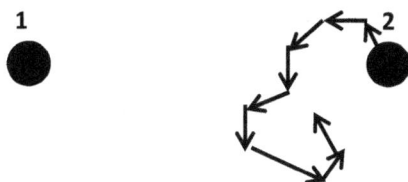

Fig. 5.5. Motion of ball 2, relative to ball 1, in a solvent.

Fig. 5.6. Motion of ball 2, relative to ball 1, in water.

solvent to be water. We find, to our surprise, that the two balls always fly toward each other (Figure 5.6). We might be puzzled by this observation since we know that there is no *direct* force exerted by one ball on the other. We also know that the solvent molecules bombard the two balls in a random way, so what is so special about water, as a solvent, in which there seems to be a new force which directs and speeds one ball toward the other. We already know that the *direct* force between the two balls is relatively weak and cannot overcome the random forces exerted by the solvent molecule. Therefore, if we do not want to invoke a "correct" direction, or a correct "target" toward which the ball flies, we must assume that it is the solvent molecules which exert a preferential directed on strong force.

In the next section, we will discuss the solvent-induced forces. We will then apply these forces to answer Levinthal's question: What are the factors that speed and direct the folding of the protein?

In the case of protein folding, it is clear that there is no target-directed pull toward the native structure. We therefore conclude that there must be a force that *forces* the protein to move along some preferential pathways. For over 50 years, scientists believed that the hydrophobic effects are dominant in protein folding (see Chapter 4). It is not uncommon to find articles and books discussing the dominance of the hydrophobic forces in protein folding, where in fact there is no discussion of any *force*, and certainly no dominant force. The fact is that the hydrophobic effect, as originally suggested by Kauzmann (1959), simply did not deliver any strong forces. In the next section, we will show that *hydrophilic forces* are much stronger than hydrophobic forces, and therefore these forces are the best candidates for answering Levinthal's question.

5.6 The Hydrophilic Forces

More than 20 years ago, strong hydrophilic forces were discovered, and implemented to explain the cause-based folding of protein [Ben-Naim (1990a, 1991a, 1992, 2009)], as well as protein–protein association and molecular recognition. These are solvent-induced forces exerted on hydrophilic groups, such as OH, C=O and NH on the protein. Once we recognize the existence of the strong hydrophilic forces, the answer to the dynamics of the PFP becomes straightforward.

There are several kinds of hydrophilic ($H\phi I$) forces. The most obvious one is the intramolecular HBs between $H\phi I$ groups of the protein. For a long time, it was believed that these HBs did not contribute significantly to the stability of the protein. In 1990, it was shown that intramolecular HBs are important, and the so-called HB inventory argument is essentially wrong (see Chapter 3). In this section, we will discuss several solvent-induced forces between $H\phi I$ groups. As we will soon see, the strongest solvent-induced forces are exerted on the various $H\phi I$ groups rather than on $H\phi O$ groups.

Before we describe the strong $H\phi I$ force, it should be emphasized that although the *force* is derived from the PMF, it is in general not true that a strong force implies strong interactions, or strong interactions imply a strong force. Figure 5.7 shows an example of a strong force derived from relatively weaker interactions.

We start with a brief *definition* of the solvent-induced effects, and then present the evidence for the relative importance of the hydrophilic forces in protein folding.

Let $G(R^M)$ be the Gibbs energy of a system of N water molecules at a given temperature T and pressure P, and

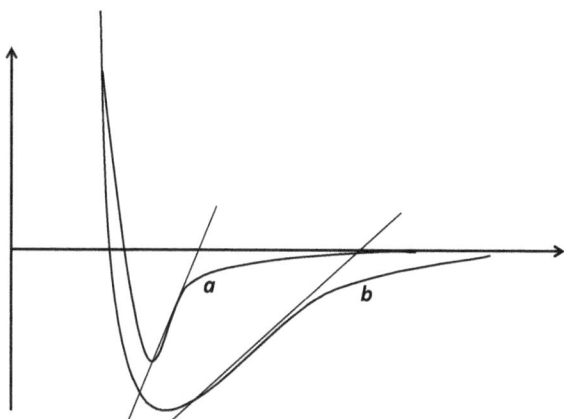

Fig. 5.7. The interaction energy of a is weaker than b. But the force in some region is stronger in a than in b.

a single protein molecule being at a specific conformation denoted $R^M = (R_1, R_2, R_3, \ldots, R_M)$, where R_i is the locational vector of atom i, or group i of the protein. It is convenient to choose the groups, such as methyl, ethyl, hydroxyl or carbonyl, rather than each atom separately to describe the conformation of the protein. The function $G(T, P, N; R^N)$ may also be referred to as the Gibbs energy landscape. In Chapter 6, we will discuss the Gibbs energy landscape in terms of the internal rotational angles.

The statistical-mechanical expression for the Gibbs energy of such a system is

$$\exp[-\beta G(R^M)] = C \int dV \exp(-\beta V)$$

$$\times \int dX^N \exp[-\beta U(R^M, X^N)]. \quad (5.1)$$

Here, $\beta = (k_B T)^{-1}$, with k_B being the Bolzmann constant and T being the absolute temperature. C is a constant having the dimensions of length to the power $-3(N+1)$. This is necessary in order to render the RHS of Eq. (5.1) dimensionless.

The quantity $U(R^M, X^N)$ in Eq. (5.1) is the total interaction energy between all the molecules in the system being at a fixed configuration. We write this as

$$U(R^M, X^N) = U(R^M) + U(X^N) + B(R^M, X^N), \quad (5.2)$$

where $U(R^M)$ is the potential energy of the protein being at a specific conformation R^M. $U(X^N)$ is the total interaction energy among all water molecules being at a fixed configuration $X^N = (X_1, X_2, \ldots, X_N)$ and $B(R^M, X^N)$ is the binding energy of the protein, i.e. the interaction between the protein at R^M and the solvent molecules at X^N.

To examine the probable direction, the protein will move in its configurational space and we need to know the *forces* acting on each of the M groups of the protein being at the conformation R^M. This force is obtained by taking the gradient of the Gibbs energy with respect to each of the R_i.

We take an infinitesimal change in R_1, keeping all other R_i ($i \neq 1$) unchanged and averaging over all the configurations of the solvent molecules. Thus,

$$-F(R_1)dR_1 = G(R_1 + dR_1) - G(R_1)$$
$$= [\nabla_1 U(R^M) + \nabla_1 \delta G(R^M)]dR_1. \quad (5.3)$$

Note that in Eq. (5.3), G is a function of R^M, but we have omitted from the notation all the R_i which do not change in this process. The first gradient on the RHS is simply the force acting on group 1 being at R_1, by all other groups of the protein. We refer to this as the *direct force*.

The second gradient on the RHS of Eq. (5.3) is the *solvent-induced force* on group 1, given a fixed conformation of the protein and averaged over all possible configurations of the solvent molecules. For any fixed conformation R^M of the protein, we can write the equality (see Figure 5.8)

$$\delta G(R^M) = \Delta G^*(R^M) - \Delta G^*(\infty), \quad (5.4)$$

where $\Delta G^*(R^M)$ is the Gibbs energy of solvation of the protein at the conformation R^M and $\Delta G^*(\infty)$ denotes the Gibbs energy of solvation of all the M groups when they are at infinite separation from each other. Since the latter is not a

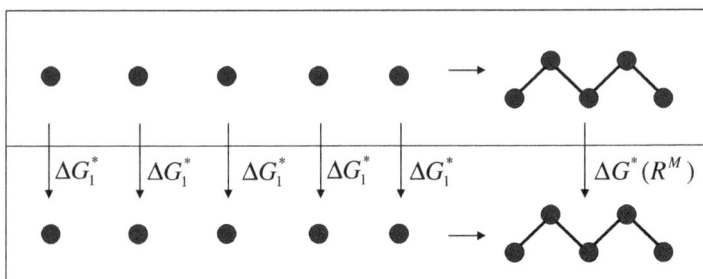

Fig. 5.8. The process of association in vacuum and in a solvent. The solvent-induced part of the Gibbs energy change is given in Eq. 5.4.

function of R_1, we can write the solvent-induced force as

$$F_1^{SI}(R_1) = -\nabla_1 \delta G(R^M) = -\nabla_1 \Delta G^*(R^M). \qquad (5.5)$$

Thus, the solvent-induced force is determined by the gradient of the *solvation Gibbs energy* of the protein at a given conformation R^M.

From Eq. (5.5), we can make the following general statement: for any given conformation R^M of the protein, a solvent-induced force will be exerted on group i, if and only if a small change in R_i causes a change in the solvation Gibbs energy of the protein [keeping T, P, N as well as $R_j (j \neq i)$ constant]. Thus, if we believe that the solvent is important in the folding process, we should focus on the indirect force in (5.5). This is exactly the component of the force which is the most relevant to the folding process.

From Eq. (5.5), it is difficult to see the various factors that contribute to the solvent-induced force. A more useful expression which is also easier to analyze and interpret is [for details, see Ben-Naim (1992, 2006)]:

$$F_1^{SI}(R_1) = \int [-\nabla_1 U(R_1, X_w)]\rho(X_w|R^M)dX_w. \quad (5.6)$$

The derivation of these equations is provided in Ben-Naim (2006, 2009). In Eq. (5.6), we see that the integrand contains two factors [note that the gradient ∇_1 operates only on the potential function $U(R_1, X_w)$ and not on $\rho(X_w|R^M)$]. The first is the force exerted on group 1 by a water molecule at $X_w = (R_W, \Omega_W)$. The second is the conditional density of water molecules at X_w given the protein at conformation R^M.

Figure 5.9 shows schematically the force $-\nabla_1 U$ (R_1, X_w), by a full line and the effect of all the groups of the protein on the density of water molecule at X_w, by a dashed line. The quantity $\rho(X_w|R^M)dX_w$ may also be interpreted as the conditional probability of finding a water molecule within the element of "volume" dX_w given the conformation R^M.

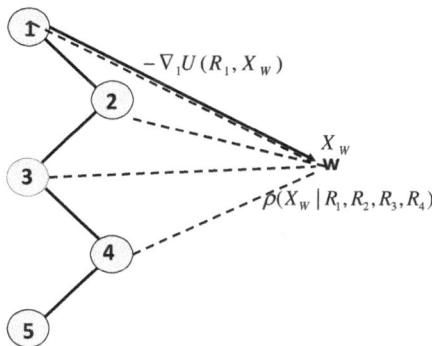

Fig. 5.9. The force exerted on group 1 by a water molecule (w), full arrow. The contributions to the local density of water molecules are indicated by the dashed lines.

Looking at the expression for the solvent-induced part (5.6), we can say that whenever there exists a strong gradient there is a strong *genuine* force. However, a strong force does not imply an instantaneous motion in the direction of the force — first, because there might be some constraints that prevent the motion in the direction of the force, and second, the solvent-induced force is statistical in nature. Therefore, all we can say is that there is a certain probability that within the timescale of our measurements, the group will move in the direction of the gradient: the stronger the force, the higher the probability of moving in that direction.

Note also that the integrand in Eq. (5.6) is a product of two factors: a *force* and a *density*. In order to have a significant contribution to the integral, both factors in the integrand must not be too small. Therefore, the integration in (5.6) is effectively over that region of space where both factors in the integrand are not negligible.

We now examine a simple case of a segment of a protein having groups of two kinds: a methyl group representing $H\phi O$ groups and a carbonyl group representing $H\phi I$ groups. In this case, we have four possible cases (Figure 5.10):

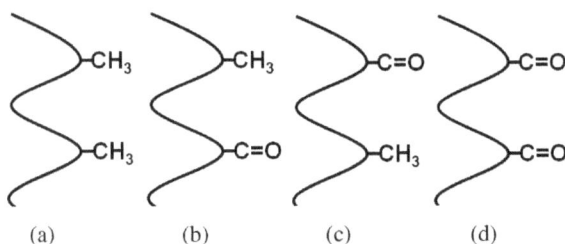

(a)   (b)   (c)   (d)

Fig. 5.10. Four cases of two groups on a protein.

(a) Group 1, $H\phi O$; group 2, $H\phi O$

In this case, the force $-\nabla_1 U(R_1, X_w)$ (between a water molecule and the $H\phi O$ group) is expected to be weak. Furthermore, since the interaction between a water molecule and the two $H\phi O$ groups is weak even at the most favorable configuration, for the triplet at $R_1 R_2$ and R_W, we expect that the conditional density at X_w will only be slightly higher than the bulk density ρ_W.

(b) Group 1, $H\phi O$; group 2, $H\phi I$

In this case, the force $-\nabla_1 U(R_1, X_w)$ is expected to be weak, as in case (a). However, because of the presence of the $H\phi I$ (group 2), there exists a distance from this $H\phi I$ group where the conditional density might be significantly enhanced. Therefore, we expect in this case to obtain a solvent-induced force larger than in case (a). Note that the larger conditional density of water molecules is expected only at a certain distance and direction from the $H\phi O$ group.

(c) Group 1, $H\phi I$; group 2, $H\phi O$

In this case, the force $-\nabla_1 U(R_1, X_w)$ is expected to be larger than in cases (a) and (b). Because of the presence of one $H\phi I$ group, the conditional density will also be enhanced. This has the same effect as the corresponding term in case (b), presuming the correct orientation of the $H\phi I$ group. Thus, in this case, we expect to get a solvent-induced force larger than in case (b).

(d) Group 1, $H\phi I$; group 2, $H\phi I$

In this case, the force $-\nabla_1 U(R_1, X_w)$ will be as large as in case (c). However, because of the presence of two $H\phi I$

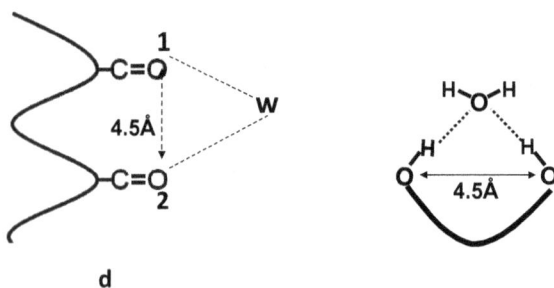

Fig. 5.11. The possibility of a water molecule forming a hydrogen bong bridge between two hydrophilic groups.

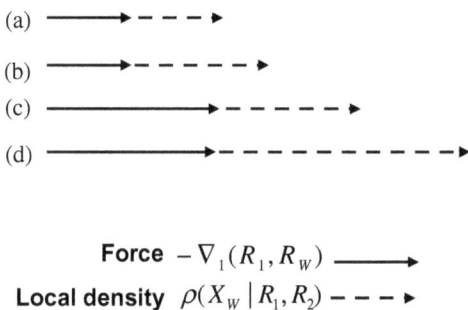

$$\text{Force } -\nabla_1(R_1, R_W) \longrightarrow$$

$$\text{Local density } \rho(X_W \mid R_1, R_2) \; \text{--- --} \blacktriangleright$$

Fig. 5.12. The relative strength of the maximal forces in cases (a) to (d) in Fig. 5.10.

groups, we might, under the right conditions, get a higher conditional density. The "right conditions" means that the two hydrophilic groups are at a distance of about 4.5 Å, and the correct orientation of the two $H\phi I$ groups (Figure 5.11). Under these right conditions, we expect large conditional density of water molecules. Thus, in this case, both the force and the conditional density will be large and the resulting solvent-induced force is expected to be larger than in case (c).

Figure 5.12 shows schematically the relative order of magnitude of the solvent-induced force from case (a) to case (d). We can conclude that the solvent-induced force on the $H\phi I$ group in case (d) will be the largest of the four cases described above. It should be noted that we have discussed the *force* exerted on group 1, in the presence of different environments. Also note that we have also examined the condition under which we expect the *maximum* forces exerted on group 1. In each case, the maximum solvent-induced force might be in different directions. We could have also taken the gradients with respect to the rotational angles ϕ_1 in the Gibbs energy function $G(T, P, N; \phi_1, \ldots, \phi_m)$. In this case, we would have obtained the torque rather than the force.

Thus, the general conclusion is that the strongest solvent-induced force is expected to be exerted on an $H\phi I$ group, when this group is also surrounded by other $H\phi I$ groups in its immediate neighborhood. We have discussed in detail only the case of *two* $H\phi I$ groups, but clearly the argument may be extended to include the effect of the presence of more $H\phi I$ groups. The force produced by such a one-water bridge is operative in the range of distances between the two $H\phi I$ groups of about 4.5 Å. Long range $H\phi I$ forces are also possible. These are discussed in Ben-Naim (1992, 2011d).

Figure 5.13 shows a possible stronger force exerted on an $H\phi I$ group (1 in the figure) in an environment of two $H\phi I$ groups. Note that the factor $-\nabla_1 U(R_1, X_w)$ is the same as in case (d). However, the conditional density is now $\rho(X_w | R_1, R_2, R_3)$. Since the three groups at R_1, R_2, R_3 are $H\phi I$, we can expect to find, under favorable locations and orientations of these groups, a higher local density of water molecules at X_w, and therefore a stronger solvent-induced force on group 1.

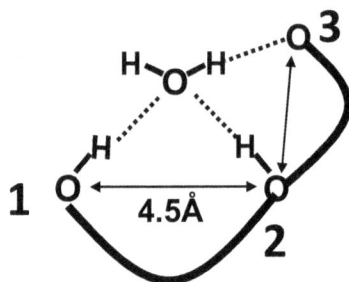

Fig. 5.13. Strong force on group 1, due to the presence of two hydrophilic groups 2 and 3.

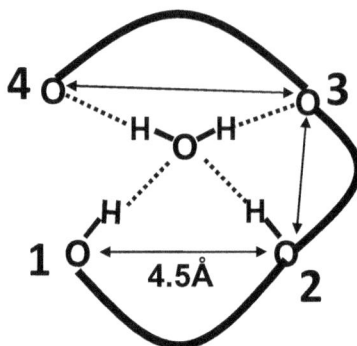

Fig. 5.14. Strong force on group 1 due to the presence of three hydrophilic groups 2, 3 and 4.

Figure 5.14 shows an example of the larger possible density. This is the case of a conditional density $\rho(X_w | R_1, R_2, R_3, R_4)$. When the locations and orientations of all the four $H\phi I$ groups are right, they can form *four* HBs with a water molecule. This will cause the largest possible local density of water molecules at X_w. Therefore, we expect to get in this case the largest possible solvent-induced force on group 1 conditioned on the most favorable $H\phi I$ environment.

So far, we have discussed *one* $H\phi I$ group. We can also consider the simultaneous force on two or three $H\phi I$ groups which are at the right locations and orientations. [For more details, see Ben-Naim (2011d).]

5.7 Implementation of $H\phi I$ Forces in Protein Folding; the Answer to Levinthal's Question

As in the case of two balls discussed in Section 5.5, if we observe a fast folding of a protein in aqueous solutions, but not in other organic liquids, then we must conclude that there are strong solvent-induced forces which both guide and speed the folding process.

As noted in the previous sections, observing a spontaneous fast-folding protein must mean that there are strong forces which cause the protein to fold in a short time. The discovery of the strong $H\phi I$ forces provides answers to both parts of Levinthal's question. The strong $H\phi I$ forces acting on the $H\phi I$ groups explain the *speed* of the folding process. The specific pattern of $H\phi I$ groups along the sequence of amino acids explains why the sequence always folds to the same structure. In other words, the pattern of the $H\phi I$ group determines the folding pathways of the protein.

It should be noted that when Levinthal made his quick estimate of the number of configurations of the protein, he assumed that each step in the folding process involved one change in one internal rotational angle. He did not consider the possibility of simultaneous change in many angles. Furthermore, as we have discussed earlier, the forces exerted on $H\phi O$ groups are weaker than the forces exerted on $H\phi I$ groups. In addition, there are about 30% side chains which are $H\phi O$. Thus, the solvent-induced forces of the $H\phi O$ groups are weak and there are not too many of these groups.

The picture changes dramatically when we consider the forces exerted on all the $H\phi I$ groups. First, each force is much larger than the corresponding $H\phi O$ forces. Second, and more importantly, there are far more $H\phi I$ groups than $H\phi O$ groups. These consist of both $H\phi I$ side chains and the C=O and NH groups of each peptide bond.

Now, imagine a protein at some unfolded configuration (either being just synthesized in the cell or having been denatured *in vitro*), with a specific sequence of forces on all the side chains and functional groups along the BB. Clearly, the specific pattern of amino acids will determines the specific *trajectory* of the protein in its configurational space. Since $H\phi I$ forces are the most important forces in the folding process, and since these forces are so plentiful, it is clear that they are responsible for the *speed* of the folding. The patterns of the carbonyl and amine groups along the BB are of course not specific to a specific sequence. Therefore, these are unlikely to determine the specific 3D structure of the protein. This could explain why there are fewer-than-expected 3D structures of the protein, and this is consistent with what Rose *et al.* (2006) suggested regarding the role of the BB in protein folding. Here, we underpin the exact reason why the BB is important — because it contains the majority of the $H\phi I$ groups.

Clearly, the BB $H\phi I$ groups cannot lead to a specific 3D structure. It is here that the $H\phi I$ side (as well as the $H\phi O$

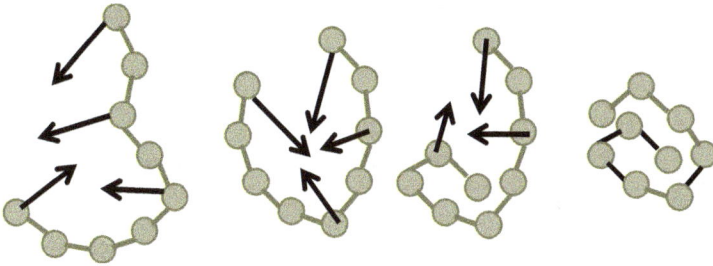

Fig. 5.15. Strong forces operating simultaneously on many groups of a protein leading to fast and guided folding.

and neutral side chains) features in determining the specific and unique pattern of forces that leads to the unique 3D structure.

To summarize, the abundance of $H\phi I$ groups on each of which a strong solvent-induced force operates leads to a simultaneous motion of multiple $H\phi I$ groups resulting in a fast and specifically directed motion of the whole protein toward the stable 3D structure. Figure 5.15 show a segment of a protein, with arrows indicating strong forces on $H\phi I$ groups.

Onc does not need to invoke the idea that the protein somehow thrives toward the *global* minimum of the Gibbs energy landscape (see Chapter 6). Certainly, one does not need to speculate on the general shape of the Gibbs energy landscape having a funnel-like or any other shape. Such speculations do not contribute anything to understanding the folding process. In my opinion, the focus on the shape of the entire GEL makes a difficult problem a far more difficult one.

It should be emphasized that in order to design a completely new protein, to execute a new function, never encountered in nature, it is not enough to design a specific sequence having a specific pattern of amino acids. One must also design the pattern of the $H\phi I$ groups at each step of the folding process. This task is almost impossible to achieve on the "drawing board."

The question is raised quite often as to how evolution has "designed" or "solved" the problem of a fast and well-guided folding of the protein. The answer to this question is "painfully" (for those who speculate how evolution did it) simple. Evolution does not *design* anything, it does not *solve* any problem, and it does not have any *target* to reach. During evolution, sequences of amino acids were synthesized much as in the analogy of the blind watchmaker. Some of these sequences happen to have a specific pattern of $H\phi I$ groups so that they happen to fold quickly.

A subset of these sequences which fold quickly happens to reach a stable 3D structure — stable enough to function in a biological system. A subset of these sequences which happens to reach a stable structure also happens to have some useful function which afforded an advantage to the biological system in which it was synthesized. Thus, the functioning proteins which we encounter and whose functions we marvel at today are not a result of any evolutionary design, but rather a result of an evolutionary process of being selected out of immense possible sequences that might have been synthesized at random.

The eventual motion of the protein toward the folded form is of course a result of both the direct and the indirect (solvent-induced) forces. Here, we were interested in the latter forces, and therefore we have focused only on the solvation Gibbs energy landscape, i.e. $\Delta G^*(\boldsymbol{R}^M)$. The gradients of the solvation Gibbs energy landscape determine the solvent-induced forces and, as we have seen above, the most important solvent-induced forces are likely to be the forces exerted on $H\phi I$ groups.

It must be stressed that the pathway of folding is never deterministic — there is always a random component to this motion in the configurational space, much like the drunken person in a big city. Once in a while, strong hydrophilic forces *force* the protein to fold in a preferential pathway. There is still a random component to this motion. However, the stronger the force the shorter the time to reach a stable 3D structure. Such a stable structure exists, not because the proteins have some inner code that commands them to reach the target, but because the protein with this particular 3D structure was selected by evolution.

It should be noted that not all the motions along the pathway should be in the direction *toward* the native 3D structure, as is implied by the target-based theories of protein folding. The protein can proceed toward or away from the native structure. What is important is that the overall pathway leads it to the final 3D structure.

Before we conclude this chapter, I would like to add a note regarding the so-called *old view* and *new view*. It seems to me that there is some confusion in the literature regarding the distinction between the kinetic pathway of the folding and the molecular, or configurational, pathway. The distinction between the two was probably first made by Baldwin (1995). In an article entitled "The Nature of Protein Folding Pathways: The Classical Versus the New

View", Baldwin writes:

The question discussed here is the nature of protein folding pathways: is there a well-defined pathway with sequential intermediates, as in an ordinary chemical reaction, or does folding follow multiple pathways without passing through a unique transition state?

This is the classical view. The new view was put forward first by theorists, based on statistical mechanical theories of folding and on Monte Carlo simulations of folding (Bryngelson and Wolynes, 1987, 1989; Shakhnovich et al., 1991; Guo et al., 1992, Leopold et al., 1992; Camacho and Thirumalai, 1993; Dill et al., 1993; Shakhnovich and Gutin, 1993; Abkevich et al., 1994a; Chan and Dill, 1994; Sali et al., 1994a,b; Shakhnovich, 1994). These studies focused on the nature of the "energy landscape" for folding, and on special properties of protein that might explain their ability to fold rapidly.

Similarly, on this distinction Rose *et al.* (2006) write:

A fundamental distinction can be made between two extreme types of systems in statistical thermodynamics, described by Baldwin as the "classical view" and the "new view." In the classical view, the behavior of the population is dictated by a small number of equilibrium states, and in this case, a general predictive theory is feasible. In the contrasting new view, the population of interest is distributed at random across a complex energy landscape, a condition that resists generalization. In the latter case, each protein will, of necessity, fold in its own unique way if, indeed, it folds at all. These contrasting concepts go to the very core of the research directions that have informed the field.

Finally, in a very recent article, Englander and Mayne (2014) show in their Figure 1 (reproduced here as Figure 5.16) the two views, and write:

The distinction between the classical view of a more or less single pathway through defined intermediates and the disordered many-pathway new view has a broad significance for the understanding of protein biophysics and biological function.

In my opinion, the "old view" is nothing but the kinetic view that the protein folds through a small sequence of intermediates, i.e. the unfolded form U goes through a few intermediates, $I_1, I_2, \ldots, I_n$, before reaching the folded form F. Each of these intermediates includes many molecular configurations. Thus, the distinction is between a macroscopic or phenomenological description of the folding pathway, and the microscopic description of the folding in terms of the motion of the protein in the Gibbs energy landscape. There is no contradiction between the two views, as is implied in Figure 5.16. The first one [Figure 5.16(a)] is a macroscopic view; the second is [Figure 5.16(b)] the microscopic view. The first is based on experiments on the kinetics of protein folding. The second is based on a hypothesis as to how protein

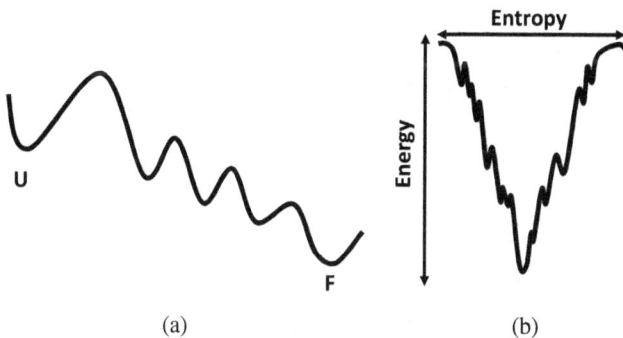

Fig. 5.16. (a) Classical view of the folding pathway. (b) The new view of multiple routes through a funnel landscape. (After Englander and Mayne (2014).)

moves in the Gibbs energy landscape. We will further discuss the latter in Chapters 6 and 7.

5.8 Conclusion

As was pointed out in the introduction, the search for a solution to the PFP as formulated by Levinthal, and as reformulated by the editors of *Science* in 2005, has gone astray twice: first, in looking for target-based solutions, and second, by adherence to the hydrophobic dominance dogma.

Once the rich repertoire of strong hydrophilic interactions as well as forces was discovered, the answer to Levinthal's question became immediate. This answer is tantamount to the solution of the dynamical part of the PFP.

Briefly, the answer to Levinthal's question is this. There are three "stages" for achieving the fast folding of the proteins. First, one must recognize that only a tiny part of the vast GEL is relevant to the protein folding. This means that there is no need to consider any kind of a random walk in the entire GEL, or assume any model for the entire GEL. Second, the strong hydrophilic forces *force* the protein to fold fast and in a specific direction [see Ben-Naim (2012b, 2013)]. Finally, the fact that there are many hydrophilic groups means that many groups will move simultaneously, under the hydrophilic forces. Thus, unlike the view that each step in the folding pathway involves one rotation about one bond, the concerted motion of the entire protein will lead to a fast folding along highly specific pathways.

In addition, the hydrophilic interactions also provide answers to the long-sought-after factors that determine the stability and specificity of the ubiquitous process of self-assembly. There remains the problem of implementation of these forces for specific proteins. This is a complicated computational problem, but it is certainly less complicated than any attempt to study the entire Gibbs energy landscape, or to speculate about the possible overall shape of the Gibbs energy landscape, vast regions of which are irrelevant to the folding of a protein, either because they are inaccessible in the normal environment (T, P, N) or because they have a very low probability of being accessed. (See also Chapter 7.)

It should be noted that any successful simulation of protein folding provides a specific answer to a *specific* protein. It can never offer a general solution to the PFP.

Recognizing the importance of the $H\phi I$ forces suggests a way to implement these forces in simulating the folding process. Having a specific protein in mind, one can scan all the forces, both direct and indirect (solvent-induced), identify the strongest or the dominant forces at that configuration and proceed to move along these forces. Repeat the scanning of the forces, find the strongest force and make the next move, and so on. This procedure will eventually lead, with high probability, to the correct 3D structure, in a relatively short period of time. There is of course no guarantee that a final 3D structure will be obtained, especially for very large proteins. Also, one should bear in mind that the result of the simulated experiment might depend on the initial conditions. This is further discussed in Chapter 6.

6

From Anfinsen's Hypothesis to the Frenetic Pursuit of the Global Minimum in the Gibbs Energy Landscape

6.1 Abstract

A major turning point in the study of protein folding was the discovery by Anfinsen of the reversible folding and unfolding of ribonuclease A. Notwithstanding the importance of Anfinsen's work, his conclusion known as the thermodynamic hypothesis (sometimes also called Anfinsen's dogma, and occasionally referred to as a thermodynamic controlled folding, i.e. the final structure is determined by the global minimum), sowed the seeds of perhaps the greatest confusion, illusions and futile efforts in the field of protein folding.

Anfinsen's thermodynamic hypothesis (quoted in the next section) was vague and ambiguous. His conclusion that the 3D structure of a protein is determined by the sequence of amino acids had led many scientists to search for a (nonexistent) folding code. The same hypothesis had led many to develop sophisticated algorithms to locate the *global minimum* in the Gibbs energy landscape (GEL), hoping that at this minimum they would find the native structure of the protein. In other words, the hope was that such an algorithm would provide a method for *predicting* the structure of the protein from the sequence of amino acids.

In this chapter, we will start from the original thermodynamic hypothesis, and show how it was misinterpreted, leading to futile efforts in searching for a *folding code*. We show that a code, in the usual sense of the word, is highly unlikely to exist. On the other hand, the search for a code, in the sense that the structure resides in the global minimum in the GEL is a result of misunderstanding the Second Law of Thermodynamics. Furthermore, the actual search for a global minimum was carried out in a complicated function which is totally irrelevant to protein. I hope that reading this chapter will discourage newcomers to the field from expending time and effort in a futile search for the native structure of proteins in that direction.

6.2 The Origin of the Myth

The origin of the idea of the existence of a folding code, and a possible hint on how to find it, were already in the writings of Anfinsen (1973):

> *The studies on the renaturation of fully denatured ribonuclease required many supporting investigations to establish, finally, the generality which we have occasionally called the "thermodynamic hypothesis." This hypothesis states that the three-dimensional structure of a native protein in its normal physiological milieu (solvent),*

pH, ionic strength, presence of other components such as metal ions, or prosthetic groups, temperature, and other, is the one in which the Gibbs free energy of the whole system is lowest; that is, that the native conformation is determined by the totality of inter-atomic interactions and hence by the amino acid sequence, in a given environment. In terms of natural selection through the "design" of the macromolecules during evolution, this idea emphasized the fact that a protein molecule only makes stable, structural sense when it exists under conditions similar to those for which it was selected, the so-called physiological value.

Before we continue, it should be emphasized that this chapter is *not* a criticism of Anfinsen's work (as some people who read a preliminary version of the chapter concluded). In fact, it has nothing to do with the importance of Anfinsen's work, or with the validity of Anfinsen's hypothesis. The criticism is directed toward those who misinterpreted, or perhaps even overinterpreted Anfinsen's hypothesis, which has led to so many unwarranted efforts in the past 40 years.

I suggest to the reader to re-read very carefully the quotation above. Try to answer the following two questions:

(1) Can you tell which *function* Anfinsen is referring to as having a minimum?
(2) Are there one or two functions?

Note first that Anfinsen is cautious in this statement, which he refers to as a *hypothesis*. Second, it says that "the native conformation is determined. . . by the amino acid sequence, in a given environment." Third, that "the three-dimensional structure of a native protein. . . is the one in which the Gibbs free energy of the whole system is lowest."

Indeed, it is true that the sequence of amino acids *determines* the structure. However, from this statement one cannot conclude that there exists a "code" which translates from "sequence" to "structure." It is also true that for a system at equilibrium the Gibbs energy of the *whole system* is at its *lowest*. From this it does not follow that the conformation of the native structure resides in the global minimum (or the lowest point) in the GEL — certainly not in the global minimum in the energy landscape (EL), or in the potential energy surface (PES).

In the following sections, we will discuss these two myths separately.

6.3 The Evolution of the "Folding Code" Myth

Soon after the publication of Anfinsen's hypothesis, many scientists concluded that there should be a folding code; see for example, Goldberg (1985). The literature was replete with statements referring to the "second translation of the *genetic message* [Kolata (1986)], "cracking the second half of the *genetic code*, "the second quarter or the second fifth" [Wolynes (2005)] of the genetic code, etc.

The very association of the *folding code* with the *genetic code* alludes to the possibility that a "code," in a sense similar to that of the *genetic code*, exists. In other words, as in the genetic code, where there is a *code* which translates from a sequence of bases (written in the four-letter alphabet of the bases) into a sequence of amino acids, people believed that there must be a code that translates from the sequence of the amino acids into a 3D structure of the native protein.

If such a code existed, it would open the way to a vast field of bioengineering. One could *design* a protein to fulfill some predetermined and desired *function*, look at the inverse code, synthesize the sequence and get the required structure. Fasman (1989) expressed that hope very eloquently:

> *The time is approaching when a new protein will be designed on the drawing board, using predictive algorithms, and its subsequent synthesis, via cloning or peptide coupling, will offer new and interesting challenges for biochemists and molecular biologists.*

In my view, this is wishful thinking. Of course, one can always *design* a new protein which is *similar* to an existent protein, to perform a *similar* function. This can be done, and in fact has been done. My reservations concern the claim that we will soon be able to design any structure we want based on the knowledge of a folding code.

My main argument, as I have explained in my recent book [Ben-Naim (2013)], is that it is not enough to *design* a sequence of amino acids which could have the required structure of a protein. One also needs to design a sequence that will *fold* into the required structure. To do so, one needs to design a sequence in such a way that at each stage of the folding process there will be a *pattern* of forces exerted on the various groups of the protein which will *guide* and *speed* the folding into the required 3D structure. This is a vastly more difficult task than designing a sequence of amino acids. This aspect of the PFP was discussed in Chapter 5.

Although I cannot offer a proof that a folding code does not exist, I will present here a few arguments which support my assertion. Recently, it became clear that the paradigm associating *function* with *structure* has to be modified. Many proteins which have important functions do not have a specific structure. These proteins are referred to as "intrinsically disordered proteins" (IDPs). It was also found that upon binding to other proteins the disordered part "folds over" the partner protein. And the structure attained depends on the partner protein.

6.4 Why a Folding Code Is Unlikely to Exist

6.4.1 *No One-to-One Code Exists*

Look again at the quotation in Section 6.2. You will see that Anfinsen says, "The three-dimensional structure... is determined... by the amino acid sequence in a given environment." The "given environment" means the temperature T, the pressure P and the composition of the solution $N = N_1, \ldots, N_C$, where N_i is the number of molecules (or moles) of the species i.

It is known that *different environments* can lead to different structures of the same sequence. The best-known example is the case of hemoglobin (Hb). Here, the different environments are different concentrations of oxygen in the solution. At a low oxygen concentration (or partial pressure), the Hb molecule attains one structure referred to as Deoxy Hb. When the oxygen concentration is higher (and hence most of the binding sites on Hb are occupied by oxygen molecules), the Hb molecule attains a different structure Oxy Hb (Figure 6.1). This fact means that a given sequence of amino acids does not *determine* a unique 3D structure.

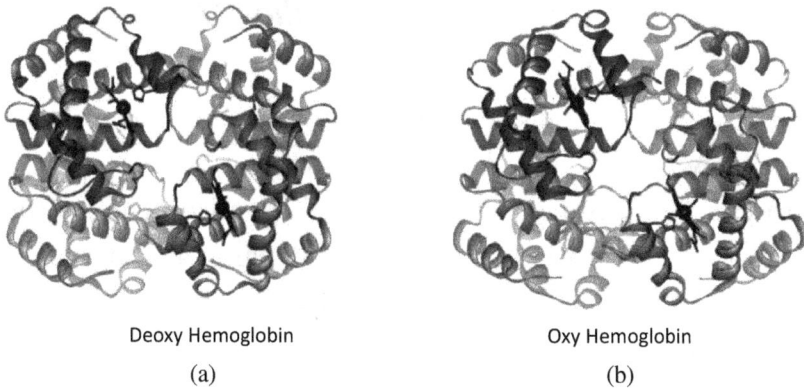

Deoxy Hemoglobin   Oxy Hemoglobin

(a)   (b)

Fig. 6.1. The two structures of hemoglobin: oxy and deoxy.

Hemoglobin is only one example in which we know that a sequence *does not* determine a unique structure. Many regulatory enzymes attain different structures according to the changing environment, which in this case is the concentration of an effector or an inhibitor [see examples in Ben-Naim (2001)]. These cases are well known. One can imagine that many different structures could be stable under different temperatures, pressures and solvent compositions. Therefore, in principle, a given sequence could lead to either no structure, one structure or many different structures, depending on the infinite possible *environments*.

On the other hand, it is well known that many sequences of amino acids can lead to the *same structure* or structures under different environments. It is therefore clear that there is no one-to-one *code* that translates from a sequence of amino acids to structures in the sense that there is a code from the DNA to protein.

6.4.2 What Is the "Alphabet" of the 3D Structure?

In the common usage of the term "code," one translates a message written in one language, expressed by one alphabet, into a second language, expressed by a second alphabet. When we talk about translating text from one language to another, the two alphabets are clear. Similarly, when we speak about the translation from DNA to proteins, the source and the target alphabets are known.

When one speaks about the folding code, one knows the alphabet of the source, i.e. the single amino acids. No one knows what the "alphabet" of the structure of the protein is. Are the "letters" of the structure the coordinates of the various amino acids in the native structure of the protein? Are the letters some rotational bond angles between amino acids [see Zwanzig *et al.* (1992)]? There were attempts to assign propensities for various amino acids to be found in different secondary structural elements of proteins. But these cannot constitute a genuine alphabet of the structure of protein.

Thus, although it is true that under a well-specified environment the sequence determines a structure (if such a structure exists at all), it does not follow that there exists a code that translates from one language (written in a known alphabet) to structure (written in unspecified alphabets, coordinates and rotational angles?).

To highlight the lack of an alphabet of the protein's structure, consider the following statement: "The sequence of bases in the DNA determines the structure and behavior of an animal." Such statements are often made in textbooks on molecular biology. In more popular books, one refers to the DNA as the "book of life," or the "blueprint" of an animal.

Clearly, not every sequence of bases determines an animal. But suppose that we restrict ourselves to well-known animals which have well-known DNA. We can say that such DNA determines the kind of animal (under some specified "environment"). From this statement, it does not follow that there is a code that translates from the sequence of bases to some characteristics (structure, behavior, etc.) of a living system. There might be some genes (a partial sequence in the DNA) which determines a specific trait of the animal (say, eye color). However, in general, it is the concerted effects of all the genes (i.e. the entire sequence of bases) which determine the eventual characteristics of an animal. In a sense, one can say that the entire animal is an *emergent* property of the sequence of DNA (in a given environment). This emergent property cannot be simply *read* from the sequence.

The idea of an emergent property — which is more than the sum of the properties of the constituents of a system — is best demonstrated by a *story* that emerges from a sequence of letters or words. Clearly, there is no code which translates from a short sequence of letters into a "short phrase" of a story. One needs to *read* and *understand* the language in which the story is written in order to "experience" the story which emerges from the sequence of letters.

In a similar sense, the sequence of amino acids determines the structure (if a structure exists). But, in order to get the structure, one needs to *read* and *understand* the language of the sequence.

The analogy between the sequence of amino acids and the letters of a book is far from perfect. In the case of protein, the *reading* of the letters is most likely done by the water molecules, and the analog of the *understanding* of the sequence would be the execution of forces exerted on the sequence of amino acids. As explained in Chapter 5, the most likely agent that executes these forces which guide and speed the folding of the sequence is the hydrophilic forces.

6.4.3 Did Evolution "Solve" the Protein Folding Problem?

In many articles and books, one finds the involvement of evolution in the PFP. Without specifying what the PFP is, one is amazed at this outstanding performance of protein, how it folds fast and to the same structure. I have heard many times the question posed during a lecture on the PFP: "How does the protein 'know' how to fold into the precise 3D structure of its native form?" Our aim — continued the lecturer — is to understand how the protein achieves this extraordinary feat.

Another version of this formulation of the PFP is to invoke "evolution" as the "agent" which has solved the PFP. Here is an example of such a statement [Wolynes (2005)]:

> Evolution solved the protein-folding problem. A major goal of biomolecular science has been to understand how this was done.

Obviously, the protein does not "know" how to fold, and there is no way we can learn from this knowledge to understand how it folds. On the other hand, it is clear that proteins were created during evolution. However, evolution did not solve the PFP, as it did not solve any problem. Evolution never *faced* a problem to be solved. Therefore, we cannot understand how evolution seemingly "solved" a problem. That is not how evolution works.

During evolution, sequences of amino acids were synthesized. A small fraction of this sequence happened to fold into some structure. Some folded more slowly, some more quickly. A small fraction of those which folded in a relatively short time happened to fold into a stable 3D structure which was relatively stable for a longer period of time. A further small fraction of sequences which happened to fold quickly into a relatively long-living 3D structure could do some useful task, or had given some advantage to the living system in which they were created.

At this moment, this tiny fraction of sequences which happened to fold quickly into a stable and functioning structure had what we call a "survival advantage." Therefore, they were "selected." The word "select," as used in daily life, does not imply a rational act of selection. The fact that something was useful, bestowing some advantage on its carrier, is the working mechanism of evolution. Evolution does not design anything, it does not solve any problem, and specifically it does not "know" how proteins fold. Thus, there is no way we can learn from evolution how protein folds.

It should be realized that the proteins we encounter today, which survived the long evolutionary process of selection, are only a tiny fraction of the unimaginable number of possible different sequences. Take a protein of, say, 150 amino acids. Since there are 20 different amino acids, there are 20^{150} possible sequences of having 150 amino acids. The vast majority of these sequences are not found as proteins in all living systems on earth. Thus, if you synthesize a sequence of, say, 150 amino acids at random, you can say with near-certainty that this sequence will not fold into some stable functioning protein. It is also clear that there will be no general method for designing proteins to perform some desired function which is very different from the existing functioning proteins.

Furthermore, it should be said that the commonly accepted paradigm that "in order to function, proteins must first fold into a precise 3D structure" will have to be changed as a result of the discovery of the so-called IDPs. Recently, many proteins which have an important function in a living system have been discovered. However, they do not have a well-defined 3D structure. The existence of IDP, also contributes to debunking the myth of the existence of a folding code. (See also Chapter 7.)

To conclude this section, we can say that a "code" in the conventional meaning of the word does not exist, and it is extremely unlikely that such a code will ever be discovered. Some people extend the meaning of a "code" to include a computer program or an algorithm which when fed with a sequence of amino acids as an input will provide us with an output of a 3D structure. Such a code does not require a code book or code words translating from one language to another. Instead, one translates the *entire* sequence into the entire 3D structure. We will devote the following sections to two such "codes." One stems from Anfinsen's hypothesis and the other is a simulation program which essentially mimics an actual experiment carried out on the computer rather than in a laboratory.

6.5 The Myth That the Native Structure Resides in the Global Minimum of the GEL

As we noted earlier, the idea of the *existence* of a folding code, and a possible hint on how to find it, were already contained in the writings of Anfinsen. In this section, we focus on that part of Anfinsen's hypothesis which led scientists to the search for the structure of a protein in the global minimum of the GEL. The relevant part of Anfinsen's article is:

> *This hypothesis states that the **three-dimensional structure** of a native protein in its normal physiological milieu (solvent), pH, ionic strength, presence of other components such as metal ions, or prosthetic groups, temperature, and other, is the one in which the Gibbs free energy of the whole system is lowest.*

Reading carefully the quotation from Anfinsen's article, we find *words* and *phrases* that "belong" to two different statements:

(i) The 3D *structure* of the native protein is at its *lowest* minimum in the GEL.

(ii) The free energy of the *whole* system is lowest.

The latter (ii) is simply a statement of the second law of thermodynamics, i.e. for a system characterized by T, P and N, the Gibbs energy attains a minimum. If that is the meaning of Anfinsen's hypothesis, then it is trivially true. However, if the first meaning (i) is accepted, then it is at best a speculation, and most likely to be wrong. Most of the people with whom I discussed this issue did not see the difference between (i) and (ii).

What did Anfinsen himself mean by his thermodynamic hypothesis? My guess is that he used the second law (ii), which is true, to imply (i). This seems to be the implication adopted by most researchers following Anfinsen, leading to the strong motivation for searching for the global minimum in the GEL. A method for finding the global minimum in the GEL would be tantamount to *predicting* the 3D structure of proteins, and this would certainly be a major achievement.

The origin of the confusion between the two statements is that both of the statements (i) and (ii) refer to the "lowest free energy." Unfortunately, both statements do not specify the *variable* with respect to which the Gibbs energy function has a minimum. This is a very subtle point which needs clarification. It should be emphasized that, whatever we will say in the following sections, it is by no means a critique of Anfinsen's work, but rather a criticism of how the thermodynamic hypothesis has been misinterpreted by scientists.

6.6 A Minimum of Gibbs Energy with Respect to What?

In mathematics, when one says that a function $y = f(x)$ has a minimum, it is clear that one refers to a minimum of the value y with respect to the variable x. The condition for such a minimum — say, at the point x_0 — is (Figure 6.2).

$$\frac{df(x)}{dx} = 0, \quad \frac{d^2f(x)}{dx^2} > 0. \tag{6.1}$$

(a)

(b)

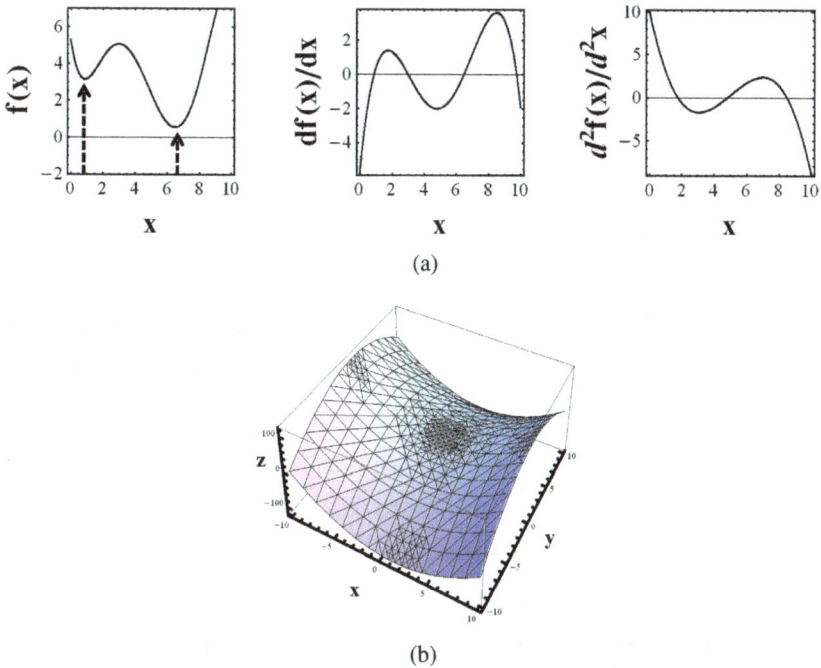

Fig. 6.2. (a) A function $f(x)$ with two minima. (b) a function $z = z(x, y)$ having a minimum in one plane and a maximum in a second plane.

A function $f(x)$ is said to have a global minimum at the point x_0 when the function of $f(x)$ has the lowest value, $f(x_0)$, in the entire range of its definition.

The same is true of a function of two or three variables: the function $v = f(x, y, z)$ having a minimum means at a point — say x_0, y_0, z_0 — when the value of v is minimal, and the criterion for such a minimum is

$$\frac{\partial f}{\partial x} = \frac{\partial f}{\partial y} = \frac{\partial f}{\partial z} = 0, \qquad (6.2)$$

$$\frac{\partial^2 f}{\partial x^2} > 0, \quad \frac{\partial^2 f}{\partial y^2} > 0, \quad \frac{\partial^2 f}{\partial z^2} > 0. \qquad (6.3)$$

Note that the minimum is always defined with respect to a small neighborhood about the point x_0, y_0, z_0. The function $f(x, y, z)$ is said to have a global minimum at (x_0, y_0, z_0) when the value of the function $f(x, y, z)$ is the *lowest* at the point (x_0, y_0, z_0). Note also that the *same point* can be a maximum with respect to one variable and a minimum with respect to another variable. For instance, the function $z = (x - 1)^2 - (y - 1)^2$ has a maximum at $x_0 = 1, y_0 = 1$ in the xz plan but a minimum in the yz plan [Figure 6.2(b)].

In thermodynamics, we say that for a system characterized by T, P, N (T is the temperature, P the pressure and N the composition of the system), the Gibbs energy G has a minimum at equilibrium. This statement is equivalent to the second law. The entropy of an isolated system characterized by constant energy E, volume V and composition N has a maximum at equilibrium.

However, unlike in mathematics, the function $S(E, V, N)$ has no maximum with respect to these variables, but with respect to some distribution functions. Similarly, the Gibbs energy function $G(T, P, N)$, does not have a minimum with respect to the variables T, P, N, but with respect to some probability distribution functions. This is, of course very abstract; in Appendix J, we will present some specific examples.

This is a serious pitfall even for mathematicians (see Appendix I).

Going back to the two possible minima (i) and (ii), we recognize that the first statement is simply a statement of stability. The 3D structure is stable in the sense that a *small* deviation from the native structure will result in an increase in the Gibbs energy of the system. Hence, there will be a thermodynamic force which will restore the conformation to its stable form. The second law does not say that this minimum must be a global minimum of the GEL. In fact, it does not say anything about the shape of the GEL.

On the other hand, the second statement is a statement of the second law. Here, the second law states that the Gibbs energy of the system at equilibrium is not only "the lowest" but also a *single* minimum (and hence a global minimum too). But the minimum is not in the GEL but a minimum of the Gibbs functional $G[\Pr(\phi)]$, i.e. a minimum with respect to all possible distribution functions $\Pr(\phi)$, where ϕ is the vector comprising all the internal rotational angles of the protein $\phi = \{\phi_1, \phi_2, \ldots, \phi_n\}$.

The two minima in the Gibbs energy function and functional are schematically described in Figure 6.3.

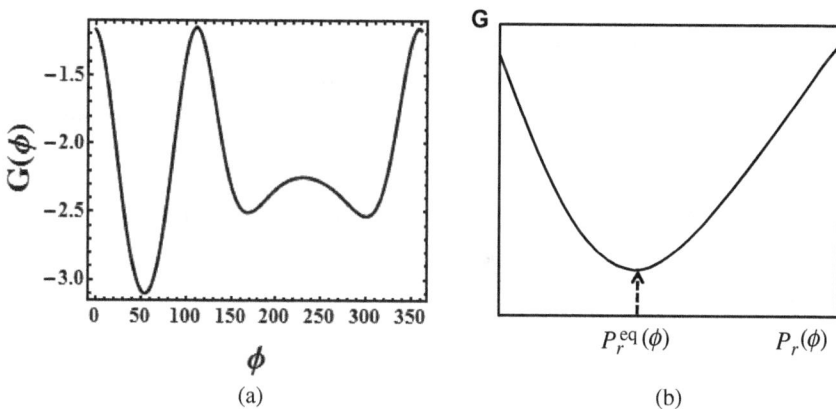

Fig. 6.3. (a) Two minima in a one-dimensional GEL. (b) A schematic depiction of single minimum of the Gibbs energy functional.

In Appendix J, we provide a few examples of minima in the Gibbs energy function and functional. In Appendix K, we provide proof that the Gibbs energy *functional* has a single (and hence global) minimum with respect to all possible *distributions*. This is essentially a statement of the second law. It is unfortunate that many scientists fell into this pitfall and interpreted Anfinsen's hypothesis as implying that the native structure of the protein *resides* in the global minimum of the GEL of a protein.

6.7 The Evolution of the Myth

As we have seen in the examples, the GEL is an extremely complicated function of all the internal rotational angles. Searching for the global minimum in such a function is an extremely difficult task.

Over the years, various sophisticated algorithms were developed. In a recent article, we find the following statements [Goldstein *et al.* (2011)].

> *It is well known that the folded state of a protein is the conformation of the lowest free energy. . . . At sufficiently low temperature the structure of the minimal free energy corresponds to the global minimum of its force field — also known as the potential energy surface (PES).*

> *The protein structure prediction problem is the problem of finding the global minimum of the PES, while the potential has a huge number of local minima.*

Then the authors continue to describe the methodology of finding the *global* minimum of the *PES*, and comparing their approach with other approaches in the literature.

Two years later, Goldtzvik *et al.* (2013) write:

Under the assumption that the molecular structure is determined by the global minimum of the potential energy surface (PES), a question that arises is how does the presence of water affect that global minimum.

Furthermore: It is well known that the folded state of a peptide (or a protein) is the conformation with the most thermodynamic stability. The global minimum of its force field (FF) — of its Potential Energy Surface (PES) — is one among an exponential set of conformations that are PES local minima.

There are three fundamental flaws in this quotation. First, the *structure* of the protein does not have to be at the global minimum of the GEL. Second, the prediction of the structure is not related to the global minimum of the GEL. Third, the PES is not the function to which the GEL tends when the temperature T is "sufficiently low" (see also Chapter 7 for the distinction between the GEL and the PES).

Indeed, "it is well known. . . , but is it *known to be true*? In other words, my concern is mainly about the question "Why should one search for a global minimum in the GEL?" In the next section, we will also comment on the claim that the native structure is at the global minimum of the PES rather than that of the GEL.

If one examines all the references provided in the articles: [Leach (2001), Schlick (2002), Finkelstein (1997), Nolting (2006), Prentiss *et al.* (2006), Finkelstein and Galzitskaya (2004), Helling *et al.* (2001), Honig (1999), Karplus (1997), Dobson and Karplus (1999), Dinner *et al.* (2000)], Goldtzvik *et al.* (2013) one will find statements similar to the above-quoted, citing references to older articles, and eventually leading to Anfinsen's thermodynamic hypothesis (1973). Since Anfinsen is considered to be the ultimate authority on this subject, not too many authors questioned the validity of his hypothesis.

The quotation above is only an example taken from the recent literature. There are many similar statements associating the 3D structure of the native protein with the *global* minimum in the GEL. A few examples are:

By definition, such a state would be at the global minimum of free energy relative to all other states accessible on that time scale.
— Dill (1990)

Optimization procedures are required for an ultimate understanding as to how interatomic interactions lead to the folded, most stable conformation of a protein from a linear polypeptide chain.

A major problem in locating the global minimum of the empirical function that describes the conformation of a protein arises from the existence of many local minima, in the multi-dimensional energy surface. — Li and Scheraga (1987)

This is a typical article dealing with the problem of searching for a global minimum of the GEL, which is a very difficult problem, but most likely irrelevant to the PFP. Furthermore, the search for the global minimum in an "empirical energy surface" is a search in a multidimensional function which is totally irrelevant to proteins.

More recently, Wales and Doye (1997) write:

It seems likely that the native structure of a protein is structurally related to the global minimum of its potential energy surface (PES). If this global minimum could be found reliably from primary amino acid sequence, this knowledge would provide new insight into the nature of protein folding, and save biochemists many hours in the laboratory.

My view is very different from that in the above quotation.

First, I do not believe that finding the global minimum of the PES will provide *any* new insight into the nature of protein folding. As to saving biochemists many hours of laboratory time, my view is that it will not only *not* save laboratory time, but add many more futile computational hours.

The purpose of this and the following sections is not to criticize any of the algorithms developed in order to find the absolute minimum in the GEL in the hope of finding where the native structure of the protein resides, hence deciphering the "folding code"; nor do I *deny* the existence of the GEL, as one author concluded from my article (see Appendix I). Instead, we will examine a more fundamental question.

Does the native conformation of the protein really reside at the global minimum of the GEL?

To answer this question, let us go back to Anfinsen's hypothesis, quoted in Section 6.1. As was explained in Section 6.5, one can think of two different meanings of "lowest free energy." One is a statement of the second law of thermodynamics. This law (formulated in terms of Gibbs energy) states that a system characterized by the variables

T, P, N has a *single minimum* with respect to all possible *distributions*. For the system of proteins in solution, we define the conformation $\phi = \phi_1, \ldots, \phi_m$ where ϕ_i is the internal rotational angle about the bond i. In association with this conformation, we define the probability density distribution $\Pr(\phi)$, where $\Pr(\phi)d\phi$ is the probability of finding the conformation between $\phi = \phi_1, \ldots, \phi_m$ and $\phi + d\phi = \phi_1 + d\phi_1, \ldots, \phi_m + d\phi_m$.

With the distribution $\Pr(\phi)$, the second law states that the *functional* $G[T, P, N; \Pr(\phi)]$ has a minimum at equilibrium over all possible distributions $\Pr(\phi)$. We denote the distributions which minimize the Gibbs energy functional by $\Pr^{(eq)}(\phi)$. Hence, the *value* of the Gibbs energy functional at equilibrium is $G[T, P, N; \Pr^{(eq)}(\phi)]$. In Appendix J, we provide a few examples for such a minimum in the Gibbs energy functional. The proof of the existence of a single distribution that minimizes the Gibbs energy functional is given in Appendix K.

Most people interpreted Anfinsen's hypothesis as implying that the equilibrium conformation of the protein is at the global minimum of the Gibbs energy *function*, i.e. of the GEL, which is a *function* $G(T, P, N; \phi)$.

The distinction between the GEL, $G(T, P, N; \phi)$ and the functional $G[T, P, N; \Pr^{(eq)}(\phi)]$ is important. The first can have zero, one or many minima. The second law *does not* state anything about where the native structure resides in the GEL. On the other hand, it states that the Gibbs energy functional has a single minimum at a *distribution* $\Pr^{(eq)}(\phi)$, to which we refer as the equilibrium distribution for some simple examples and the proof of the existence of the minimum in the Gibbs energy functional; see Appendices J and K.

The confusion between the GEL and the Gibbs energy function is partially notational, and partially a result of misunderstanding the second law. One can easily remove the confusion resulting from the notation by defining the GEL as

$$G = f(T, P, N; \boldsymbol{\phi}) \tag{6.4}$$

and the Gibbs energy functional as

$$G = F[T, P, N; \Pr(\boldsymbol{\phi})]. \tag{6.5}$$

Thus, f is a *function* of the conformation $\boldsymbol{\phi}$ but F is a *functional* of the density distribution function $\Pr(\boldsymbol{\phi})$. At equilibrium, the GEL and the equilibrium distribution are related to each other by

$$\Pr^{(eq)}(\boldsymbol{\phi}) = C \exp[-\beta f(T, P, N; \boldsymbol{\phi})]. \tag{6.6}$$

This relationship applies for fixed values of T, P, N.

The notational confusion results when we write

$$G = G(T, P, N; G(T, P, N : \boldsymbol{\phi})). \tag{6.7}$$

Here, the G on the LHS is the *value* of the Gibbs energy. The other two G's are the *names* of the *function* and the *functional*.

Thus, if the distribution function $\Pr^{(eq)}(\boldsymbol{\phi})$ has a sharp maximum at the native conformation $\boldsymbol{\phi}_N$, we can say that the GEL has also a sharp minimum at the native conformation. However, one cannot conclude that in general this minimum is the global minimum in the GEL. For small polypeptides, it might be the case that the global minimum is also the location of the most stable conformation. However, for large proteins it is extremely unlikely that the native structure resides in the global minimum in the GEL. This certainly is not a result of the second law of thermodynamics.

Furthermore, we will see that most people who search for the global minimum in the GEL are actually searching for a minimum in a function (PES) which is not even related to the GEL and totally irrelevant to the PFP.

6.8 The Search for a Global Minimum in the EL or in the Potential Energy Surface

In the earlier literature, we find that people looked for the global minimum in the EL rather than in the GEL. This is sometimes referred to as the PES. More recently, the PES has been constructed based on the EL, with corrections due to solvent effects; see below. Note that the EL is relevant to a protein in vacuum. It is sometimes argued that at $T \to 0$ the ground state must be occupied by all the molecules. This is true in the case where all the conformations are accessible. On the other hand, in solutions the GEL is the relevant quantity. In this case, the argument that at $T \to 0$ the protein must attain the global minimum does not apply.

There is another confusion between the EL and the thermodynamic *internal energy* of the system, which is denoted by $\Delta U(i \to f)$. The latter is approximately equal to $\Delta H(i \to f)$, but not to $\Delta E(i \to f)$; see below. These correspond to two different splittings of the Gibbs energy changes.

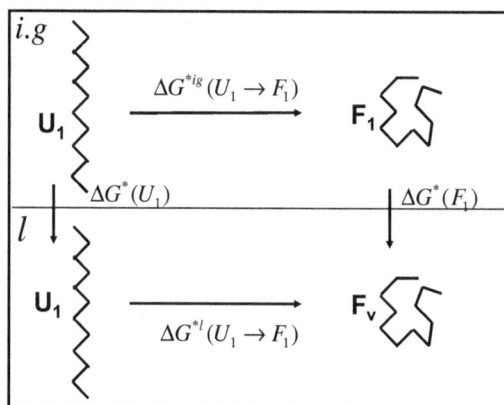

Fig. 6.4. The relationship between Gibbs energy of the folding process and Gibbs energy of solvation of the U and F forms.

Suppose that we perform the process of conversion from one initial conformation (i) at a fixed position in the liquid to another final conformation (f), as in Figure 6.4. The Gibbs energy change for the process is

$$\Delta G(i \rightarrow f) = \Delta E(i \rightarrow f) + \Delta G_f^* - \Delta G_i^*. \qquad (6.8)$$

Here, in the gaseous phase, the change in the Gibbs energy for the process ($i \rightarrow f$) is the energy change $\Delta E(i \rightarrow f)$. On the other hand, we have the thermodynamic identity

$$\Delta G(i \rightarrow f) = \Delta H(i \rightarrow f) - T\Delta S(i \rightarrow f), \qquad (6.9)$$

where ΔH and ΔS are the enthalpy and the entropy changes of the process ($i \rightarrow f$) in the solution. These can be respectively written as

$$\Delta H(i \rightarrow f) = \Delta E(i \rightarrow f) + \Delta H_f^* - \Delta H_i^*, \qquad (6.10)$$
$$\Delta S(i \rightarrow f) = \Delta S_f^* - \Delta S_i^*. \qquad (6.11)$$

Since the process $(i \to f)$ is from a fixed conformation (i) to a fixed conformation (f), the change in the enthalpy in vacuum is $\Delta E(i \to f)$, and the change in the entropy in vacuum is zero.

Therefore, from (6.9), (6.16) and (6.11), it follows that for $T \to 0$

$$\Delta G(i \to f) \to \Delta E(i \to f) + \Delta H_f^* - \Delta H_i^*. \quad (6.12)$$

Thus, it is not true that as $T \to 0$ the GEL tends to $\Delta E(i \to f)$. Even at $T \to 0$, the GEL differs from the EL by the solvation enthalpies of the conformers i and f.

In the early literature one can find searches for the global minimum in the EL. This is obviously a function irrelevant to proteins in solutions. More recently, people have recognized that the relevant function is the GEL. Unfortunately, we are very far from having an idea of what this landscape looks like.

It should be said that while we have a reasonably good idea of the various contributions to the EL, we are totally in the dark with respect to the GEL of even a modestly sized protein. In some cases, one starts with the EL and "corrects" it to obtain an approximation to the GEL, which is referred to as the PES. Thus, the approximate GEL is written as

$$\text{GEL} \approx \text{EL} + \sum_{i,j} \delta G_{ij}^{H\phi O}, \quad (6.13)$$

where $\delta G_{ij}^{H\phi O}$ stands for pairwise hydrophobic $(H\phi O)$ interactions between groups i and j. Unfortunately, such a corrected EL cannot be used as an approximated GEL: first, because in (6.13), one takes only $H\phi O$ interactions, and neglects the more important (both stronger and more abundant) hydrophilic interactions; second, and not less important, because the solvent-induced part of the GEL is not pairwise-additive (Appendix L). Thus, in effect, one ends up searching for a global minimum in a very complex function which is neither an EL nor a GEL, and not relevant to protein folding at all.

6.9 The Search for the Structure of Protein by Simulation

The second route to obtaining the structure of the protein is simply by brute force simulation on the computer. If one has a good model for the protein, for water, and for the interactions between water molecules and the protein, there is in principle no difficulty simulating the folding of a protein, and getting a stable 3D structure, if such a structure exists. Such a simulation program is sometimes referred to as a computer code. However, this kind of code is not a code in the usual sense of the word. In fact, a simulation of protein folding is more akin to an experiment of folding rather than a theoretical "prediction" of the structure of a protein.

Clearly, the larger the protein is, the more difficult and time-consuming the simulation calculation will be. For protein, the structure of which is known, there should be no problem in starting the simulation from the known structure, then denaturing it, and renaturating it.

However, for very large proteins there might be a serious difficulty in finding the initial conditions for the simulation which will lead to the stable 3D structure even when such a structure exists. In addition, since the eventual structure depends on the environment (P, T, N), one might miss the correct structure due to the fact that the model used for water is not perfect.

The hindrance to reaching the 3D structure is sometimes referred to as kinetic controlled folding. Basically, it means that on the way to the 3D structure the protein might encounter a potential barrier which might hinder its smooth flow toward the stable structure. However, with very large proteins there might be such high potential barriers that completely obstruct the protein's pathway to the 3D structure. This is a very fundamental difference between a folding of a natural protein into its stable 3D structure, and the simulated folding of the same protein on a computer.

To clarify this point, we will first discuss some general features of the GEL for very large proteins. We will then discuss two types of experiments: Anfinsen-type and non-Anfinsen-type. The distinction between these two types is important when one is contemplating the simulation of protein folding for a very large protein.

6.9.1 The Anfinsen-Type Experiment

Suppose that we start with the native conformation of a known protein in some environment (T, P, N). We then change the environment (say, by changing the temperature, the pressure or the composition of the solvent), and obtain denaturation of the protein. Next, we restore the original environment T, P, N, and the protein's conformation is restored to the original native conformation. This is essentially Anfinsen's experiment.

Of course, not all proteins can be renatured — say, by restoring the original temperature. In this respect, Anfinsen was "lucky" to choose a protein which was renatured by restoring the original conditions.

On a molecular level, we can characterize the Anfinsen-type experiment as follows. We start with the native protein in some initial *environment*, T, P, N. In this environment, there is some distribution of the states of the protein which we denote $\mathrm{Pr}^{(\mathrm{in})}(\phi)$, where ϕ is the conformation of the protein. It consists of all the internal rotational angles of the protein.

Next, we change the environment to say, T^1, P, N (i.e. we raise the temperature from T to T^1), and we obtain a new distribution of states which characterizes the denatured state. Let us denote this distribution $\mathrm{Pr}^{(\mathrm{den})}(\phi)$.

Finally, we restore the initial condition to T, P, N, and require that the initial distribution also be restored. In order for this to happen, it is necessary that each of the states in the natured protein will be *accessible* from each state of the denatured protein.

This scenario is very different from the current view where the renaturation is described as a flow, or as funneling on the GEL, eventually leading to the global minimum in the GEL. We will further discuss the funnel metaphor in Chapter 7. Here, we emphasize the erroneous interpretation of Anfinsen's hypothesis in terms of "flow" toward the global minimum of the GEL.

Of course, we cannot always guarantee that the protein will return to its original structure under the restoration of the initial environment. We will examine in this section the conditions under which we can achieve a renaturation of the

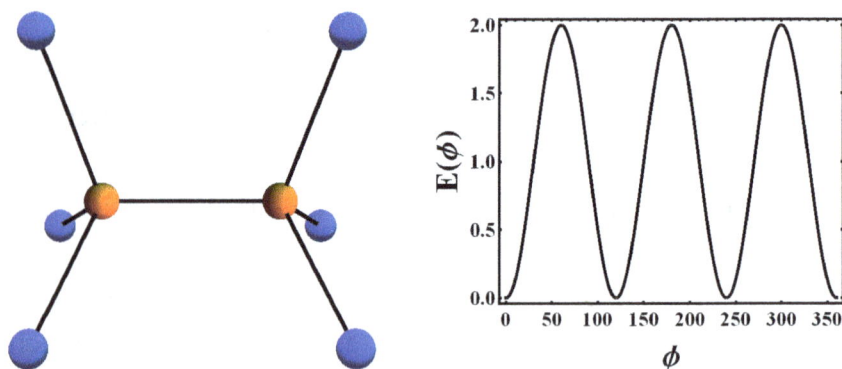

Fig. 6.5. The general shape of the EL of ethane.

protein. In the next subsection, we will discuss the case of a non-Anfinsen-type experiment.

For the illustrations in this and the next section, we use a one-parameter EL and GEL. The simplest system is a rotation about the C–C bond in ethane. The EL in this case is a function having three minima and three maxima. In the case of an ethane molecule in solution, the GEL is defined by (see Figure 6.5)

$$G(\phi) = E(\phi) + \delta G(\phi). \qquad (6.14)$$

In this particular case, we do not expect that the shape of the GEL will differ appreciably from that of the EL. Note that in this section, we use the notation $G(\phi)$ instead of $G(T, P, N; \phi)$ for the GEL.

Next, we use a substituted ethane molecule which has two carboxylic acids. We assume that at some range of conformation the solute molecules can form two HBs with a water molecule. The corresponding GEL is shown schematically in Figure 6.6.

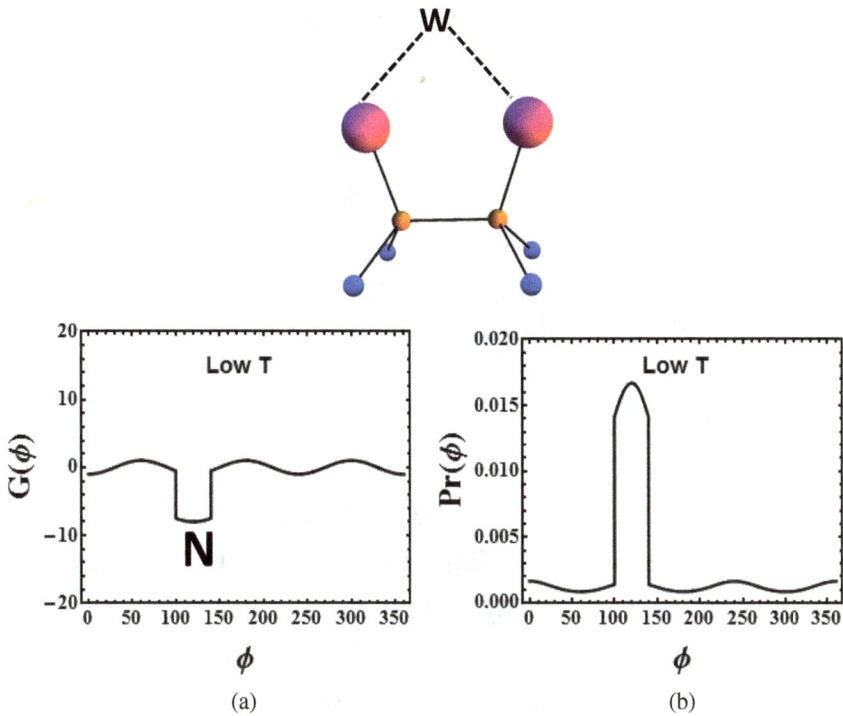

Fig. 6.6. Possible shape of the GEL and the probability distribution for a substituted ethane molecule at a low temperature.

This particular shape was calculated in a way similar to the computations of the exact GEL in 1D systems [Ben-Naim (2011b)]. The details of the computations are unimportant. We use Figure 6.6 as well as the next figures for illustration only. Here, we see that at an angle of about 120°, there is a minimum in the GEL.

Thus, if the GEL of the solute is as in Figure 6.6, then we expect that most of the solute molecules will reside in the region of conformation $100° \leq \phi \leq 140°$. We call these conformations the native state N. In Figure 6.6, we draw the GEL and the corresponding probability distribution

(ϕ), which is derived from the GEL by

$$\Pr(\phi) = \exp\frac{[-\beta G(\phi)]}{C}, \qquad (6.15)$$

$$C = \int_0^{360} \exp[-\beta G(\phi)]d(\phi), \qquad (6.16)$$

where C is the normalization constant. Thus, in this case, the probability distribution will have the same form as in [Figure 6.6(b)].

Starting from the system at "low T" as in Figure 6.6, we now perform the Anfinsen-type experiment. We first increase the temperature to a "high T." At this temperature, the GEL will have the same shape as in [Figure 6.7(a)], and the corresponding probability distribution as in [Figure 6.7(b)]. As can be expected, in this particular state there is no single region in which most of the molecules reside. There is a finite probability of finding the solute molecule at any of the conformations $0° \leq \phi \leq 360°$.

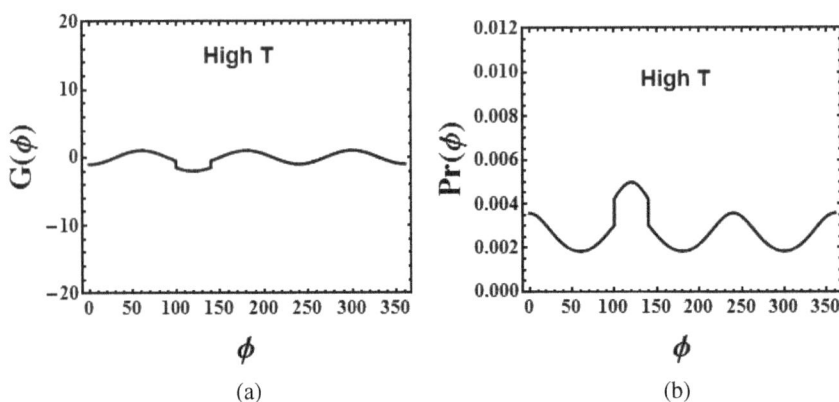

Fig. 6.7. Possible shape of the GEL and the probability distribution for a substituted ethane molecule at a high temperature.

We now restore the original temperature (low T). The second law of thermodynamics requires that the probability distribution shown in [Figure 6.7(b)] change to the unique probability distribution of the environment "low T." In other words, the Gibbs energy has a *single minimum* with respect to *all possible probability distributions*. The second law *does not* require that the *conformations* of the molecules "flow" into any particular state. Nevertheless, in this particular example, we see that *most* of the molecules will attain a conformation in the region denoted N (native) in [Figure 6.6(a)].

It should be emphasized that the apparent "flow" of most of the conformation into the region N is not a result of the existence of a global minimum in the GEL. In this particular case, we chose the GEL in such a way that it has a global minimum. However, the minimum of the Gibbs energy required by the second law is valid with respect to the *entire probability density function*, or the entire GEL, as shown in [Figure 6.6(a)]. In the next section, we show an example for which the native conformation does not necessarily have to be at the global minimum of the GEL.

Before doing this, we illustrate another Anfinsen-type experiment involving change in the environment — in this case, the concentration of adding a denaturation agent. In the simple Anfinsen experiment, one adds a denaturing solute — say, urea — which causes denaturation of the protein; then one removes the denaturing solute, and restores the original native structure. One can apply exactly the same line of reasoning as we did for the temperature effect, using Figures 6.6 and 6.7, but reinterpreting the high T and the low T as a high concentration or a low concentration of, say, urea, respectively.

Here, we discuss another case of the effect of adding a solute (i.e. changing the environment) which does not cause denaturation, but instead transforms the solute from one stable conformational state to another. A well-known example is hemoglobin, which has two stable structures, oxy and deoxy hemoglobin, depending on the concentration of the oxygen in the "environment."

Hemoglobin is not the only example of such a solute-induced structural transformation. Many regulatory enzymes operate by changing their structure in response to the changes in the concentration of an activator or inhibitor molecule. This fact is relevant to the question of predicting the structure of the protein given the sequence. We will demonstrate the situation with a simple example.

Suppose that in an environment with a low concentration of oxygen, the GEL has the same shape as in [Figure 6.8(a)].

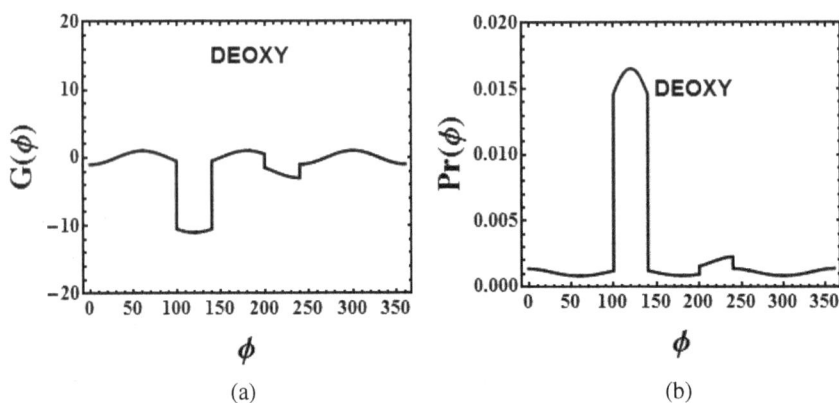

(a)   (b)

Fig. 6.8. Possible shape of the GEL and the probability distribution for substituted ethane molecule at a low oxygen concentration.

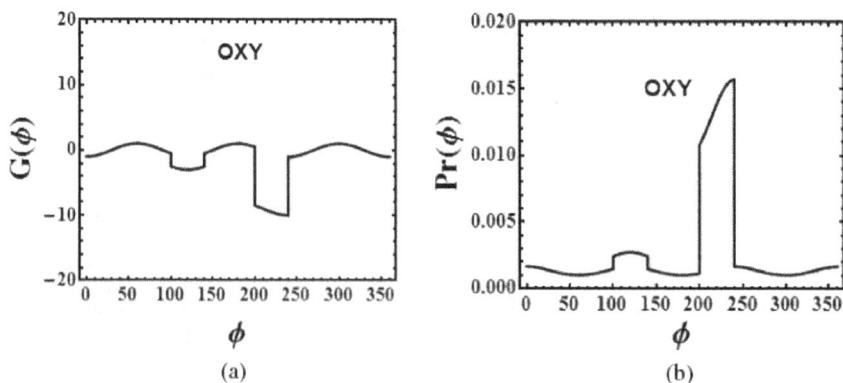

Fig. 6.9. Possible shape of the GEL and the probability distribution for a substituted ethane molecule at a high oxygen concentration.

This means that at equilibrium most of the molecules will be in the region $100° \leq \phi \leq 150°$. We refer to this conformation as "deoxy." On the other hand, when we increase the oxygen concentration, the GEL of the system changes into a new shape, as in [Figure 6.9(a)]. Under these conditions, the majority of the molecules will be at one of the conformations $200° \leq \phi \leq 250°$, which we refer to as the "oxy" conformation.

It is sometimes said that the GEL of such a system has two deep minima, at the two stable conformations. The more correct statement is that the system under *different* concentrations of oxygen has two *different* GELs, each of which has one dominating conformation corresponding to one deep minimum.

When we start with a system with a low oxygen concentration, we are in the GEL of [Figure 6.8(a)]. In this environment, there is an equilibrium distribution of species [Figure 6.8(b)]. When we add oxygen, the molecules do not "flow" from one minimum in the GEL to another minimum in the same GEL. Instead, in the new environment (high concentrations of oxygen), the GEL that minimizes the Gibbs energy functional is *different* from the GEL that minimizes the Gibbs energy functional in the new environment, and this *new* GEL happens to have a minimum at a different location. Alternatively, we may say that by adding oxygen there is a new equilibrium distribution of species [Figure 6.9(b)].

6.9.2 The Non-Anfinsen Type of Renaturation Experiment

In this subsection, we discuss an example of a non-Anfinsen type of renaturation experiment. This type of experiment sounds like it is an exception to the Anfinsen-type experiment. In fact, we will argue that the contrary is true. Namely, the Anfinsen type is more likely to be the exception, especially for *larger polypeptides*. We will also discuss the relevance of these types of experiment to the question of predictability (or, rather unpredictability) of the structure of the protein, given its sequence, as well as their implication for simulated experiments on protein folding.

A non-Anfinsen type of experiment may be characterized experimentally as follows. We start with a protein in its native structure. Then we denatured it by changing its environment. When we restore the original environment, either the protein does not return to its original native structure or only a part of it returns to its original structure.

On a molecular level, we can characterize a non-Anfinsen experiment in terms of the accessibility of the native state. Let

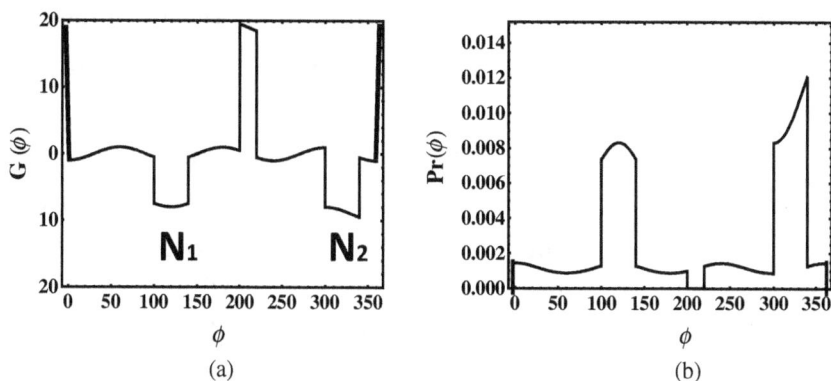

Fig. 6.10.   A possible GEL having two very high Gibbs energy barriers.

$Pr^{(in)}(\phi)$ be the original distribution of states and $Pr^{(den)}(\phi)$ the distribution of the denatured protein.

If some or all of the original states are inaccessible from the denatured states, then, upon restoration of the original environment, the protein will not return to its original distribution of states.

We will clarify the situation with a simple example.

Figure 6.10 shows a schematic example of a GEL at a low temperature. The main feature of this GEL is that it contains two regions which are separated by very high potential barriers at $\phi = 0°$ and about $200°$. This means that there are two "isomers," or two regions of states which are inaccessible from each other.

Now, suppose that the native structure is located at the minimum at about $120°$ (N_1 in [Figure 6.10(a)]). Note that the distribution of states of the protein is mostly within the region $100°-150°$. In the Anfinsen-type experiment, we heat the system and the probability distribution becomes flatter, but still there is no transition between the two regions separated by the high potential barrier. Therefore, when we cool the system all the molecules will attain the original distribution.

In our example, when we heat the system, some of the molecules can cross over the potential barrier at 200°. At this temperature, the molecules will be distributed in the entire range of conformations $0° \leq \phi \leq 360°$. When the system is cooled to the original temperature, it will not attain the distribution defined by

$$\Pr(\phi) = \frac{\exp[-\beta G(\phi)]}{\int_0^{360} \exp[-\beta G(\phi)]d\phi}. \qquad (6.17)$$

This distribution is shown in [Figure 6.10(b)]. Instead, some of the molecules whose states are to the left of the barrier at 200° will now be distributed as in [Figure 6.11(a)], with a high probability of being at $\phi \approx 130°$. On the other hand, those molecules whose states are to the right of the barrier at 200° will now be trapped in the region $220° \leq \phi \leq 350$, with a high probability at $\phi \approx 330°$ [Figure 6.11(b)].

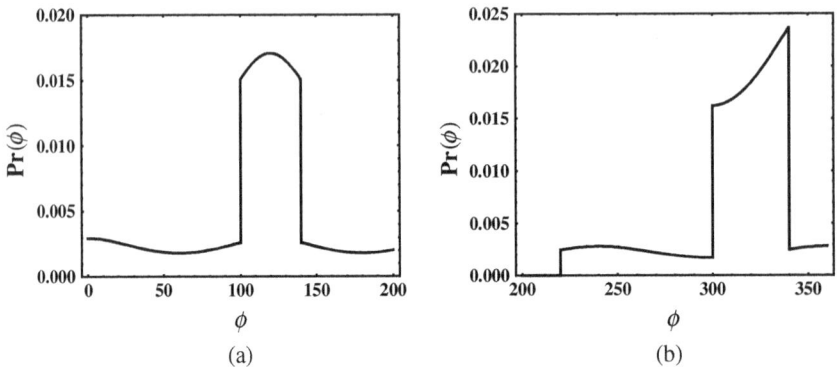

Fig. 6.11. Two distributions of conformations which correspond to the two disconnected regions in Fig. 6.10.

We see that, in this case, restoring the original temperature will not restore the system to the original distribution of states. Thus, if the native state is at N_1, then after we restore the original temperature the molecules will split into two isomers: one with a distribution as in Figure 6.11(a), and the other with a distribution as in Figure 6.11(b). The same is true if we start with the native state at N_2, and heat it to the same high temperature as in the previous case.

Note that in this example we have a GEL with two well-defined minima. This does not need to be the general case. For instance, in Figure 6.12, we show a GEL in which the native state (N) is at about 50°. In this case, after denaturation and renaturation, most of the molecules will occupy the region $140° \leq \phi \leq 320°$, and perhaps only a few will attain the original native state at $\phi \approx 50°$.

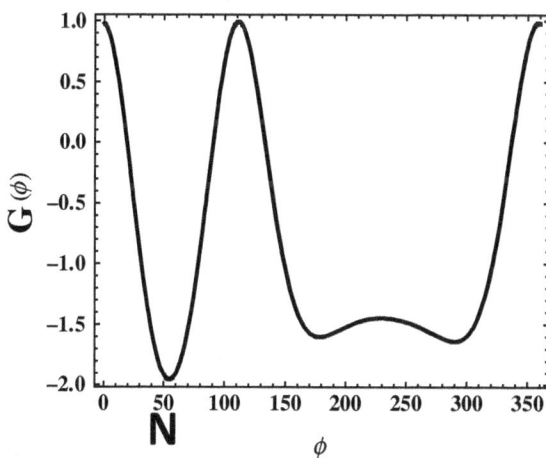

Fig. 6.12. A possible GEL with one deep minimum. After denaturation, part of the molecules will occupy the region to the right of the maximum at about 120°.

6.9.3 *Dependence on the Initial Conditions*

The discussion of the examples above has a profound implication for simulation experiments on the folding of a large protein. In such experiments, one starts with an initial distribution of states — say, $Pr^{(in)}(\phi)$ — in some initial environment. Unlike the denaturation–renaturation experiment, where the environment is changed and restored, here the environment is kept constant. Therefore, the outcome of the simulation should depend on the initial distribution of states, and on the accessibility to other regions in the GEL in the *same* environment.

Before we present a few examples, it is important to emphasize that the GEL is defined over the *entire configurational space*, including all states which are inaccessible. However, the probability distribution in that environment is in general not the one given by Eq. (6.17).

We turn to a few examples.

Suppose that we start with an initial distribution which is a Dirac delta function $\delta(\phi - \phi_1)$ [Figure 6.13(a)], where $\phi_1 \approx 50°$. Suppose that at the same T, P, N the GEL is as shown in [Figure 6.10(a)], where there is a very high potential barrier at $\phi \approx 200°$. In this case, by releasing the constraint on the initial distribution of species, the system will relax *not* to the probability distribution defined in Eq. (6.17) and shown in [Figure 6.10(b)], but only to that region of conformations which are accessible from ϕ_1 at the given T, P, N. This is shown in [Figure 6.13(b)]. In other words, the new distribution function will be

$$Pr(\phi) = \frac{\exp[-\beta G(\phi)]}{\int_0^{200} \exp[-\beta G(\phi)]d\phi}, \qquad (6.18)$$

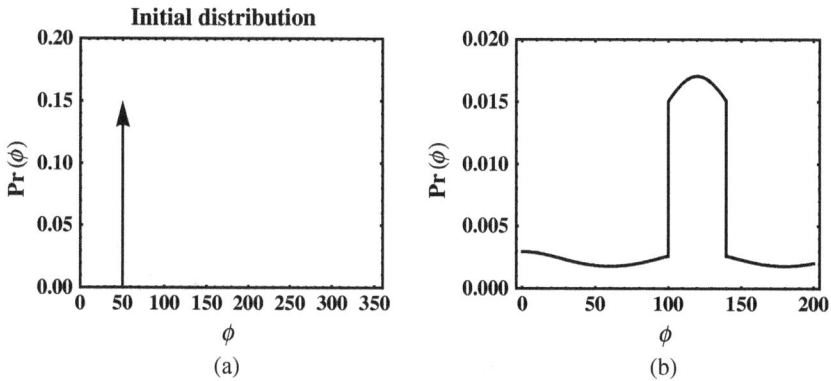

Fig. 6.13. An example of (a) an initial distribution, and (b) relaxing to a final distribution.

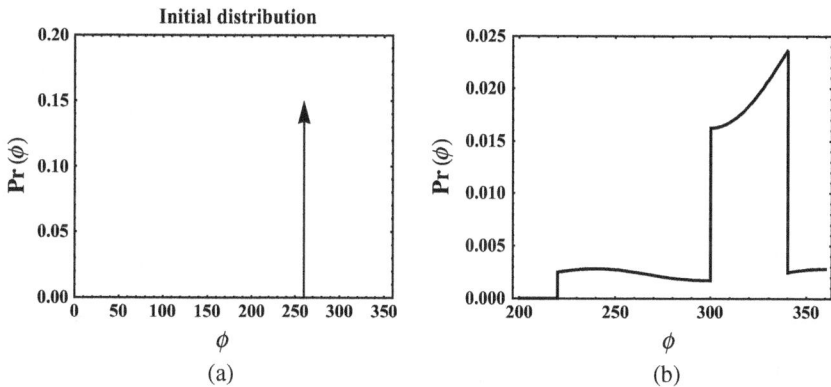

Fig. 6.14. An example of (a) an initial distribution, and (b) relaxing to a final distribution.

where now the normalization constant is an integral over the accessible region from ϕ_1.

A second example is shown in [Figure 6.14(a)]. Here, we start with an initial distribution $\delta(\phi - \phi_2)$, where $\phi_2 \approx 260°$. In this case, when the constraint on the initial distribution is released, the system will relax to the distribution which is

defined by

$$\Pr(\phi) = \frac{\exp[-\beta G\phi]}{\int_{200}^{360} \exp[-\beta G\phi]d\phi}. \qquad (6.19)$$

This is shown schematically in [Figure 6.14(b)]. Note that in each case, we maintain a constant environment, and barriers at $\phi = 0°$ and $\phi = 200°$ are not crossed.

Finally, if we start with any arbitrary initial distribution as shown in [Figure 6.15(a)], and release the constraint on the fixed distribution, the system will relax to *two* different new distributions. All the species for which $\phi \lesssim 200$ will

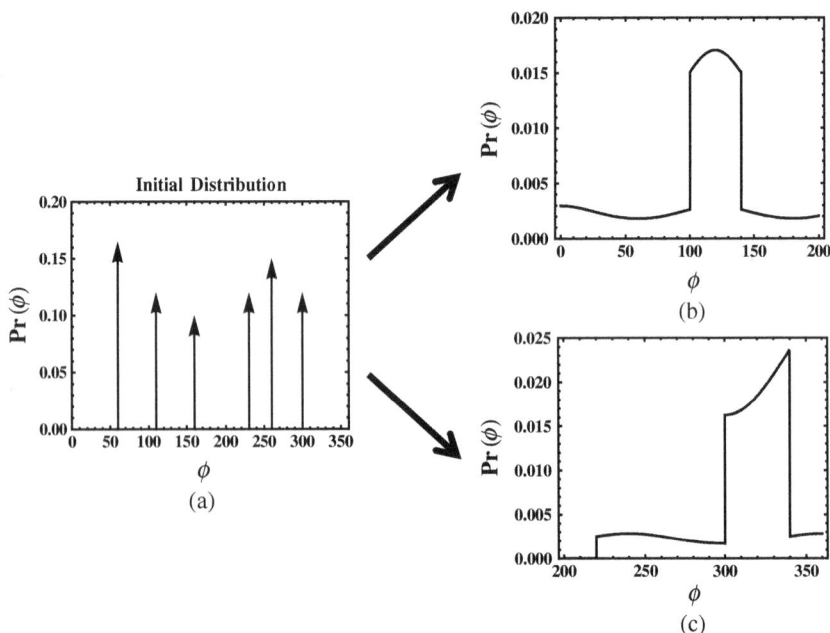

Fig. 6.15. An example of (a) an initial distribution, (b) partially relaxing to a final distribution, and (c) partially relaxing to a different distribution.

relax to a new distribution, defined in Eq. (6.18) and shown schematically in [Figure 6.15(b)]. On the other hand, all the species for which $\phi \gtrsim 200$ will relax to a different distribution, defined in Eq. (6.19) and shown schematically in [Figure 6.15(c)].

In all the three examples discussed above, the conformations of the solute will not "flow" into the global minimum of the GEL. Instead, in each case the species will flow into a probability distribution of states which are accessible from the initial distribution.

Note that even if we start with all molecules at the conformation corresponding to the global minimum of the GEL and heat the system, or add a denaturation solute, there is no guarantee that by restoring the original conditions the system will flow back into the global minimum of the GEL. In the Anfinsen-type case, the original probability distribution will be recovered, but in the non-Anfinsen-type case it may or may not be.

The examples discussed in the previous sections were relevant to a particular simple solute, the conformational state of which is characterized by one rotational angle, ϕ. However, the conclusion reached above is valid for any protein and any number of internal rotational angles.

Before we discuss the general case, we need to introduce the following notation. The GEL is defined on the *entire* region of variation of all the rotational angles. In the one-parameter case discussed in previous sections, the entire domain on which the GEL is defined by

$$D_{\text{total}} = \{0 \le \phi \le 2\pi\}, \tag{6.20}$$

For the general protein, the GEL is defined on the domain D_{total}, which includes the *entire* range of variation of each rotational angle ϕ_i between 0 and 2π.

We have seen that in the one-parameter case, the total domain may be split into two disconnected subdomains at some given environment (T, P, N). By "disconnected subdomains," we mean that at the given T, P, N, there exist high potential barriers between the two domains so that at equilibrium the conformations accessible from any specific conformation are only within one domain. All other conformations included in other domains are inaccessible.

In a two-parameter case, we expect to find more than two disconnected subdomains. Figure 6.16(b) shows the

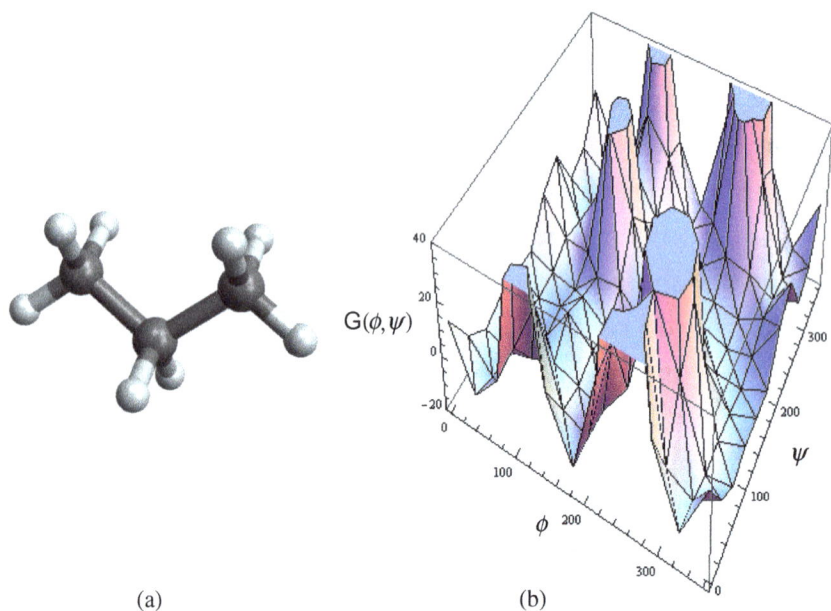

(a)   (b)

Fig. 6.16. Possible GEL for a molecule having two internal rotations, e.g. a substituted propane molecule.

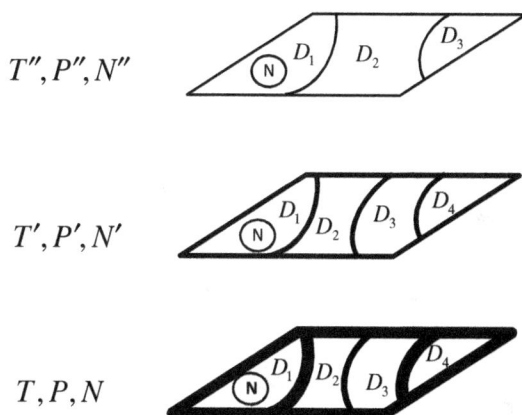

Fig. 6.17. Schematic disconnected regions in the GEL, separated by an insurmountable barrier.

GEL of a two-dimensional case. The specific curve in the figure was obtained by substituting bulky spheres to propane [Figure 6.16(a)]. In a real protein with about 150 amino acids, we expect to have many more disconnected domains.

In this GEL, there are many regions which are completely inaccessible in an ordinary environment. These regions are shown by heavy lines in Figure 6.17. In addition, some of these lines could demark the borders between different and disconnected domains.

Note that the connectivity between the various subdomains is dependent on the environment. Figure 6.17 shows that both the inaccessible regions, as well as the number of disconnected domains, can change with the temperature. As we have noted above, the GEL includes *all* possible conformations of the protein, including those that have a zero probability of occurrence, as well as all the regions that are disconnected. The latter are mainly due to the possibility of forming knots (Figure 6.18). In large proteins, we expect

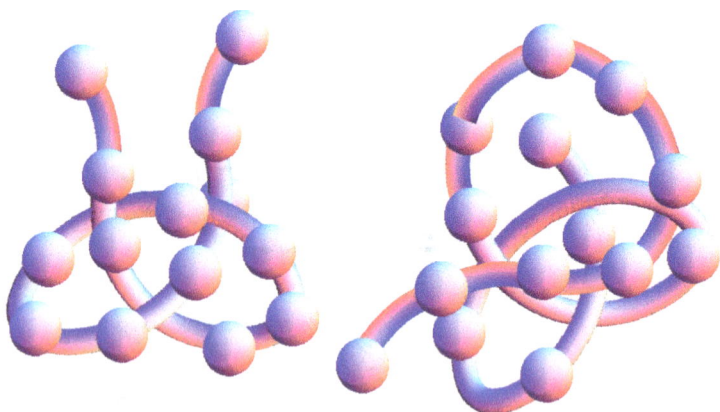

Fig. 6.18. One and two knots in a polymer.

many possible knots; the configurations of all these are *included* in the GEL. In a given environment, one cannot expect that any subregion will be accessible from any another subregion.

In the Anfinsen-type experiment, we start with the native structure. Let us denote by $D_N(T, P, N)$ the domain of the native structure at a given T, P, N. Also, let us assume that D_N is included in one of the domains — say, D_1 — which is disconnected from all other domains at T, P, N (Figure 6.19). We now change the environment to, say, T', P', N', and we obtain a new distribution of conformations $x^{eq}(T', P', N')$. We denote by $D_U(T', P', N')$ the domain of the conformations of the unfolded protein. Note that $D_U(T', P', N')$ is, in general, a very small part of the entire GEL, i.e. of D_{total}. Now, we restore the original environment to (T, P, N). If most of the species in the distribution $x^{eq}(T', P', N')$ belong to the domain that includes the domain D_1 (which includes the native domain D_N), then

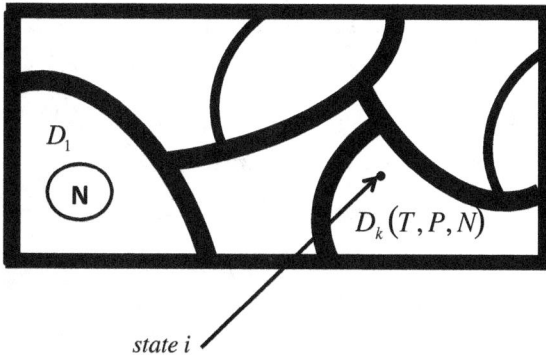

Fig. 6.19. An example of an initial state which belongs to one specific domain D_k.

the system's Gibbs energy will reach a minimum at the original conformational distribution $x^{eq}(T, P, N)$. This is the Anfinsen-type experiment.

On the other hand, if the distribution $x^{eq}(T', P', N')$ contains species in domains which in the original environment T, P, N were disconnected, then there is no guarantee that the protein will flow to the original distribution $\mathbf{x}^{eq}(T, P, N)$. This is the case of the non-Anfinsen experiment. The same conclusion is valid for a simulated experiment. If we start from any arbitrary initial distribution at T, P, N, there is no guarantee that the system will flow into the native domain $D_N(T, P, N)$ even if such a native conformation exists.

Regarding the relative sizes of the various domains, we know that in a given environment T, P, N, the native structure holds a very small domain, $D_N(T, P, N) \ll D_{total}$. Note that D_{total} is *independent* of the environment. As we have seen before, the *total* domain of the GEL is likely to contain a very large number of disconnected subdomains. The size of these domains is of course environment-dependent.

Thus, for each environment (T, P, N), the total domain D_{total} is split into many disconnected subdomains, $D_1, D_2, \ldots, D_l$ (we ignore here regions in the GEL which have a zero probability of occurrence), where

$$\bigcup_{k=1}^{l} D_k(T, P, N) = D_{total}. \qquad (6.21)$$

In each domain D_k, there exists an equilibrium distribution of species $x^{eq}(T, P, N; k)$. Thus, if we start with a single conformation i, which belongs to a specific subdomain $D_k(T, P, N)$ in a given environment (see Figure 6.19), and then relax the constraint on the fixed conformation i, the free energy of the system will be lower — not because the initial conformation will flow to the native conformation, but because the system will attain an equilibrium distribution of conformations which belong to the domain D_k, i.e. the domain which includes the initial conformation i.

By generalization, if we start with any arbitrary initial distribution of conformations $x^{in} = (x_1^{in}, \ldots, x_m^{in})$ (either obtained experimentally or chosen arbitrarily in a simulated experiment), there is no guarantee that the conformations will flow to the native conformation (if such a conformation exists). Some of the conformations will relax to the equilibrium distribution on the domain $D_1(T, P, N)$, some others to a distribution on the domain $D_2(T, P, N)$, and so on. It is only in the Anfinsen-type experiment when all (or most) of the initial conformations belong to the same domain, which includes the domain of the native conformation, that the system will relax to the original equilibrium distribution belonging to that domain.

If we start with an *arbitrary* initial conformation in a simulated experiment with a very large polypeptide, and let the system relax to equilibrium, it is guaranteed that the system will relax to the equilibrium distribution in that domain which includes the initial conformation. It is extremely unlikely, however, that the system will relax to the equilibrium distribution which includes the *native* conformation, even if such a native conformation exists as that environment.

Thus, the endeavor of searching for the native structure of a large protein at the global minimum in the GEL is not only unjustified theoretically, but even if one finds the global minimum in the GEL, it is extremely unlikely that this minimum will coincide with the native conformation. On the other hand, when one simulates protein folding of a large protein, the resulting structure (or no structure) will depend strongly on the initial distribution of states.

6.10 Conclusion

We started this chapter with a quotation of Anfinsen's hypothesis. The hypothesis refers to the "lowest Gibbs energy," which, if understood correctly, is nothing but a statement of the second law of thermodynamics (for a system at T, P, N constants).

Most scientists interpreted Anfinsen's hypothesis as implying that the native structure of the protein resides at the global minimum of the GEL. Such an interpretation is a result of a misunderstanding of what the GEL is, as well as what the second law states. It is unfortunate that much effort has been expended in searching for the global minimum of the GEL, in the hope that the native structure of the protein resides there.

The search for the 3D structure of the native protein at the global minimum of the GEL is oftentimes described metaphorically as "looking for a needle in a haystack." The conclusion of this chapter is better described as "looking for a needle in one out of many haystacks, in which the needle might not even reside." On the other hand, the search for the 3D structure in the PES can be likened to a search in the wrong haystack, i.e. the search for the global minimum of the PES is as futile as searching for a minimum in a function which is irrelevant to proteins.

The analysis made in this chapter has also some relevance to simulation of the folding process of either a protein of known structure or a completely new polypeptide. In the first case, the result of the simulated experiment would depend on the initial conditions. If one starts from any arbitrary initial conformation, it is extremely unlikely that the resulting equilibrium distribution will contain the native structure. On the other hand, for a new polypeptide we cannot know in advance in which environment (T, P, N) this polypeptide will fold, if it folds, to any stable 3D structure.

Thus, the conclusion of this chapter is threefold. First, it is extremely unlikely that a folding code exists. Second, the search for the "prediction" of the structure of protein in the global minimum of the GEL is a result of misunderstanding the second law of thermodynamics. Finally, the search for the global minimum of either the EL or the PES is a search in a very complicated function which is irrelevant to proteins in a solvent.

7

Some Candidates Which Can Potentially Evolve into New Myths

In this chapter, I will mention a few terms, phrases and concepts which have been featuring recently in the literature, and which I believe have the potential to evolve into new myths. Most of these are relatively new in the field of protein folding, and as of now cannot be referred to as being established myths, paradigms or dogmas. The myths discussed in the previous chapters are well established and well cited in the literature. They still linger in the literature despite the overwhelming evidence debunking them. The new topics discussed here do not yet enjoy widespread acceptance. In fact, many scientists who are active in the field and with whom I have talked, dismissed them as being empty words and having no scientific content.

The views expressed in this chapter are my own. They were influenced by many scientists, with whom I have talked or had e-mail exchanges. The scope and the content of this chapter are left open-ended. I would be glad to receive any comments or criticisms, any suggestions to add potential myths or remove some of those discussed here. Hopefully, all these will be included in a new edition of this book.

7.1 The Landscape Theory of Protein Folding

Before I start to discuss the "landscape theory" of protein folding, a few clarifying notes are in order. There seems to be considerable confusion between various "landscapes." The most important ones are the energy landscape (EL), Gibbs energy landscape (GEL), thermodynamic potential energy landscape (PEL) and potential energy surface (PES).

Most people who discuss the "landscape theory" do not bother to *define* what they mean by "landscape." Some do not care to specify which landscape they are referring to, and to my best knowledge no one has clarified what the "theory" in the "landscape theory" is.

In most cases, people explicitly refer to EL theory meaning GEL theory, but actually describe the PEL theory. Then they draw a funnel-like shape which is supposed to represent the shape of the EL, GEL or PES. We will further discuss the funnel idea in Section 7.2.

All these are *functions* of all the coordinates which are used to describe the coordinates or the conformations of a protein. In some publications, the "landscape" is referred to as the Hamiltonian of the system. This is even more confusing, since the Hamiltonian normally also includes the kinetic energy of the molecule. The landscape usually involves only the potential energy part of the Hamiltonian.

7.1.1 *The Energy Landscape*

The EL is the potential energy of a *single* protein as a function of the locational coordinates of either all the atoms or all the internal rotational angles. In the first case, we can define the EL as the change in the potential energy for the process of bringing all the m atoms of the protein from infinite separation to the final configuration R^M [Figure 7.1(a)]. Thus,

$$\Delta E = E(R_1, \ldots, R_m) - E\left(R^M = \infty\right), \qquad (7.1)$$

where $R^M = R_1, \ldots, R_m$ is the configuration of the entire molecule and R_i is the locational vector of the center of the ith atom. The configuration $R^M = \infty$ refers to all atoms at infinite separation from each other. Since $E(R^M = \infty)$ is chosen as zero, the EL can be simply defined as $E(R^M)$. Note that $E(R^M)$ includes all the chemical

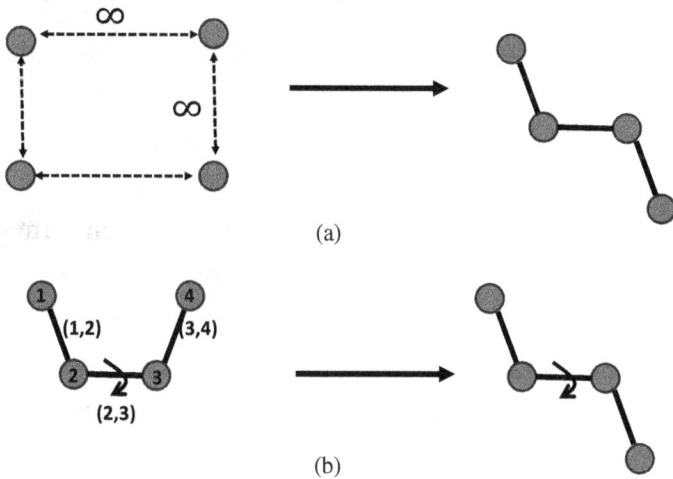

Fig. 7.1. Two possible definition of the energy landscape; see details in Section 7.1.1.

bond energies within the molecules. In proteins, we usually assume that the bond energies are unchanged upon changing the configuration of the molecule. In this case, it is more convenient to define the EL as a function of all the internal rotational angles, $\boldsymbol{\phi}^m = \phi_1, \ldots, \phi_n$. One can choose any reference state, $\phi_1^0, \ldots, \phi_n^0$, to define the energy change:

$$\Delta E \left(\boldsymbol{\phi}^0 \to \boldsymbol{\phi}\right) = E(\phi_1, \ldots, \phi_n) - E\left(\phi_1^0, \ldots, \phi_n^0\right). \quad (7.2)$$

Since $E(\phi_1^0, \ldots, \phi_n^0)$ is a constant, we can refer to $E(\phi_1, \ldots, \phi_n)$ as the EL of the protein.

Figure 7.1 shows a schematic definition of the two functions $E(\boldsymbol{R}^M)$ and $E(\boldsymbol{\phi}^n)$ of butane.

It should be noted that EL is a *function*, not a theory. Also, in general, one cannot assign to a single protein any thermodynamic function, like entropy or Gibbs energy. The energy function is a well-defined quantity. One can also define the Shannon measure of information of a protein, presuming that one knows the probability distribution $\Pr(\boldsymbol{\phi}^n)$ of all the conformations.

The term "landscape" attached to this energy function is nothing but a visual description of the function. Like the landscape one sees out of the window, one can imagine a multidimensional landscape. It is a function of n variables. In the simplest case of one rotational angle, the landscape is a one-dimensional function as shown in Figure 7.2, and for a molecule with two angles, it is a two-dimensional function as shown in Figure 7.3. In both of these plots, there is no room for the "entropy" of the molecule (see the next section on the funnel metaphor for the EL).

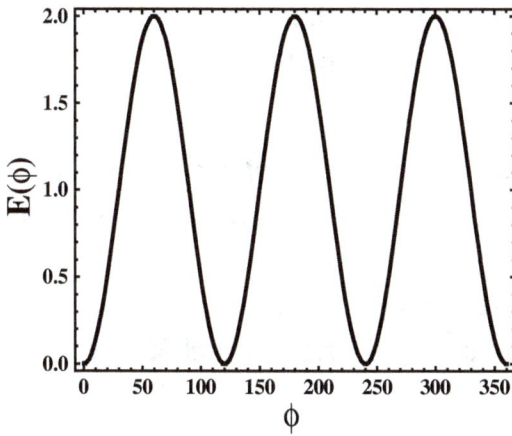

Fig. 7.2. A simple one-dimensional energy landscape.

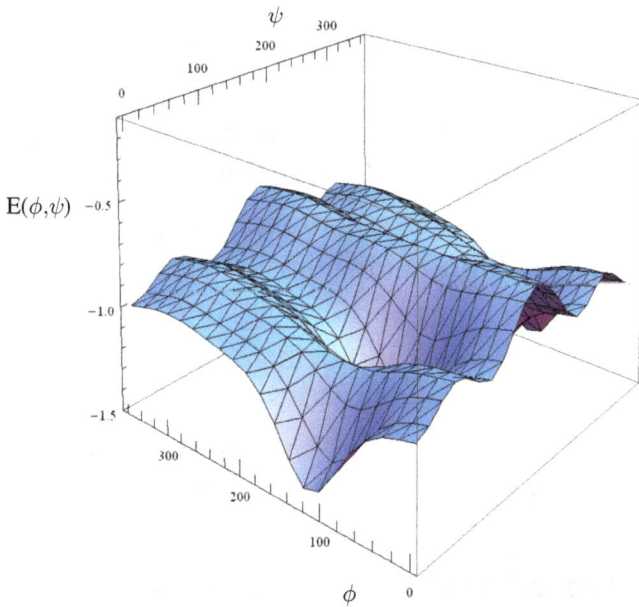

Fig. 7.3. A simple two-dimensional energy landscape.

A few properties of the EL should be noted, especially when we compare it to the GEL:

(1) The domain on which the function $E(R^M)$ or $E(\phi^n)$ is defined includes *all configurations*. These also include configurations for which the energy is infinitely large. In polymer chemistry, such configurations are referred to as excluded volume. They occur whenever two atoms or groups of atoms overlap in such a way that the repulsive energy is infinitely high. Figure 7.4 shows a hypothetical one-dimensional landscape derived from ethane in which two hydrogen atoms are replaced by bulky atoms [in this illustration, we simply replace two hydrogen atoms by two Lennard–Jones (LJ) spheres]. As we increase the diameter of the LJ sphere, we reach a point where the two spheres penetrate into each other, and hence the potential function becomes infinitely large. Such configurations are not *realizable* and are usually excluded from the discussion of the conformations of the polymer. However, the EL by definition includes all of these configurations, i.e. the entire range of the angles $0 \leq \phi_i \leq 2\pi$, for all i.

(2) The EL, to a good approximation may be assumed to be pairwise-additive. In the case of $E(R^M)$, the energy function includes all the interactions between all pairs of atoms, including all chemical bonds. On the other hand, in $E(\phi^n)$ the energy function does not include the chemical bonds. It is only a function of the internal rotational angles ϕ_i, keeping all bond distances and bond angles unchanged. Thus, in butane the EL, $E(R^M)$ includes all pairs of interactions [Figure 7.5(a)], whereas

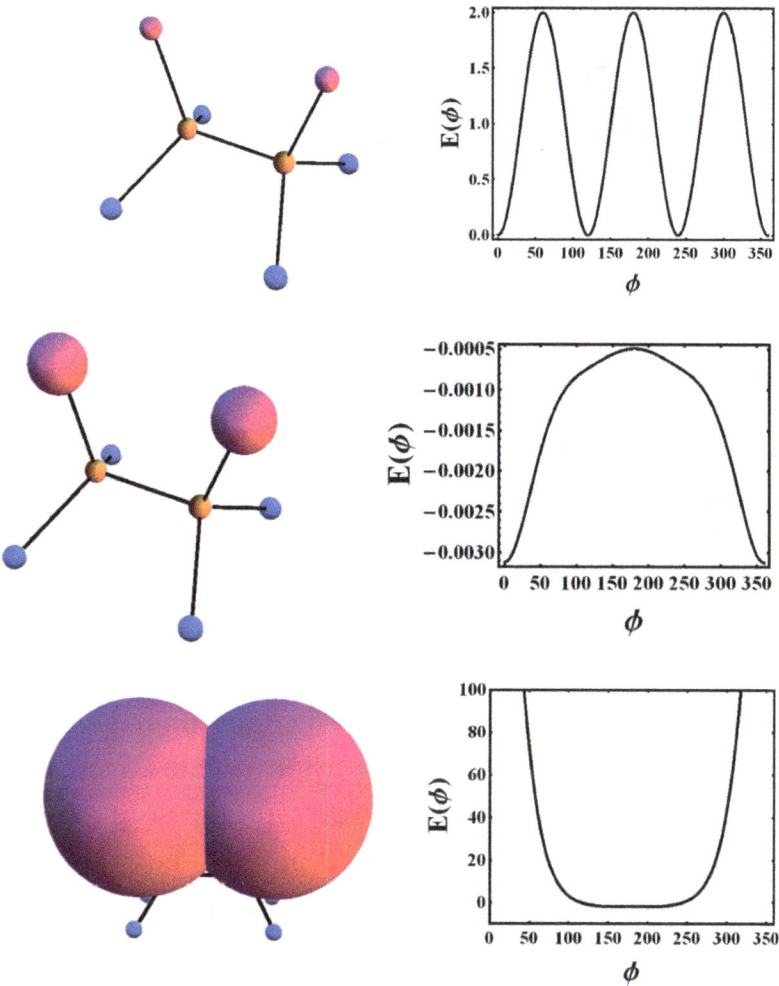

Fig. 7.4. More complicated one-dimensional energy landscape.

in $E(\phi^n)$ is the change in the potential energy when the molecule rotates about the 2–3 bonds, i.e. this is a one-dimensional function.

We note here that this kind of pairwise additivity is lost when we proceed from the EL to the GEL.

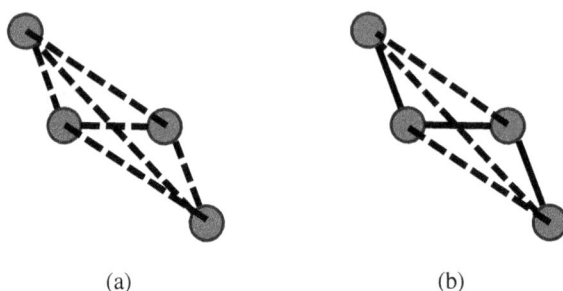

(a)   (b)

Fig. 7.5. The interaction energies (dashed lines) between (a) all atoms, or groups, and (b) only nonbonded atoms or groups.

(3) The EL, to a good approximation, is presumed to be independent of the temperature, pressure or chemical composition in the environment. Again, this property is lost when we proceed to the GEL.

(4) Finally, we note that for a protein of, say, 200 amino acids, the EL is a fairly complicated function. However, we have a good idea of the various ingredients that contribute to the EL, as well as approximate values of their magnitude. Because of the additivity approximation, one can build up the function EL from the knowledge of its components — a knowledge obtained either experimentally or theoretically from small compounds.

We can safely say that when we proceed from the EL to the GEL the complexity of the function is many orders of magnitude larger. (See next section.)

7.1.2 *The Gibbs Energy Landscape*

The best way to define the GEL is as follows.

Suppose again that we choose some standard conformation ϕ^0 and define the process $\phi^0 \to \phi$, having the

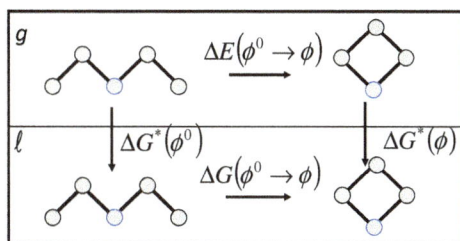

Fig. 7.6. The relationship between energy change, Gibbs energy change and the solvation quantities.

corresponding change in energy

$$\Delta E = E(\boldsymbol{\phi}) - E\left(\boldsymbol{\phi}^0\right), \qquad (7.3)$$

Now, we insert the molecule in a solvent as shown in Figure 7.6. Note that in this figure, the polymer is at a fixed conformation (either $\boldsymbol{\phi}^0$ or $\boldsymbol{\phi}$) and has a fixed location and orientation in the gaseous and liquid phases.

From Figure 7.6, one can write the following equation for the change in the Gibbs energy for the same process $\boldsymbol{\phi}^0 \to \boldsymbol{\phi}$, but now carried out in a solvent at some given temperature T, pressure P and composition $N = (N_1, \ldots, N_c)$:

$$\Delta G\left(\boldsymbol{\phi}^0 \to \boldsymbol{\phi}\right) = \Delta E\left(\boldsymbol{\phi}^0 \to \boldsymbol{\phi}\right)$$
$$+ \Delta G^*(\boldsymbol{\phi}) - \Delta G^*\left(\boldsymbol{\phi}^0\right), \quad (7.4)$$

where $\Delta G^*(\boldsymbol{\phi})$ and $\Delta G^*(\boldsymbol{\phi}^0)$ are the solvation Gibbs energies of the molecule at $\boldsymbol{\phi}$ and ϕ^0, respectively. (See Appendix A for the definition.)

We can also rewrite Eq. (7.4) as

$$\Delta G\left(\boldsymbol{\phi}^0 \to \boldsymbol{\phi}\right) = \Delta E\left(\boldsymbol{\phi}^0 \to \boldsymbol{\phi}\right) + \delta G\left(\boldsymbol{\phi}^0 \to \boldsymbol{\phi}\right), \quad (7.5)$$

where δG is defined as the solvent-induced change in the Gibbs energy of the system in the process $\phi^0 \to \phi$:

$$\delta G\left(\phi^0 \to \phi\right) = \Delta G^*\left(\phi\right) - \Delta G^*\left(\phi^0\right). \qquad (7.6)$$

Since $\Delta G^*(\phi^0)$ is a constant, we can define the GEL either by Eq. (7.4) or simply as

$$G\left(\phi^0 \to \phi\right) = E(\phi) + \Delta G^*(\phi). \qquad (7.7)$$

It should be noted, however, that in writing $G = G(\phi)$, G stands for both the *name* of the *function* (here the GEL), and the *value* of the Gibbs energy. This notation may cause some confusion when we want to distinguish between the GEL and the Gibbs energy functional. (See Chapter 6 for details.) In most articles discussing the "energy landscape theory" of protein folding, the authors frequently confuse EL and GEL, but almost always do not specify which function is referred to when discussing the GEL.

The notation $G(\phi)$ is a shorthand for the function $G(T, P, N; \phi)$, which is the change in Gibbs energy for the process $\phi^0 \to \phi$ carried out in a solvent at a constant T, P, N. For simplicity, we will usually omit T, P, N from the notation.

The function $G(\phi)$ is the Gibbs energy of a macroscopic system at T, P, N having one protein at a fixed position, orientation and conformation. In reality, we never deal with a single molecule in a solvent at a fixed position, orientation and conformation. However, it can be shown that in the limit of a very dilute solution of the protein in a solvent, the Gibbs energy of the entire system at T, P, N, N_p, where N_p is the number of protein molecules [all at fixed positions (but far from each other), orientations and conformations], can be expanded as

$$G(T, P, N, N_p; \boldsymbol{\phi}) = G(T, P, N) + N_p \mu_p^*(\boldsymbol{\phi}), \qquad (7.8)$$

where $\mu_p^*(\boldsymbol{\phi})$ is the pseudochemical potential of the solute at $\boldsymbol{\phi}$. This is defined as the change in the Gibbs energy of a system at T, P, N due to the insertion of *one* solute into the solvent at a fixed location, orientation and conformation. If the system is sufficiently dilute, such that there are no interactions between solute molecules, then the contribution of the solute molecules to the Gibbs energy of the system is simply $N_p \mu_p^*(\boldsymbol{\phi})$, i.e. $\mu_p^*(\boldsymbol{\phi})$ per solute molecule.

If the system contains N_p solute molecules at a specific conformation $\boldsymbol{\phi}$, but free to move and rotate in the solution, Eq. (7.8) should be replaced by

$$G(T, P, N, N_p) = G(T, P, N) + N_p \mu_p(\boldsymbol{\phi}), \qquad (7.9)$$

where $\mu_p(\boldsymbol{\phi})$ is the chemical potential of the solute at conformation $\boldsymbol{\phi}$. The relationship between μ_p and μ_p^* is

$$\mu_p(\boldsymbol{\phi}) = \mu_p^*(\boldsymbol{\phi}) + k_B T \ln \rho_p \Lambda_p^3, \qquad (7.10)$$

where k_B is the Boltzmann constant, ρ_p the density of the protein ($\rho_p = N_p/V$) and Λ_p^3 the momentum partition function of the protein. [For more details, see Ben-Naim (2006, 2009).]

In passing from a fixed position and orientation to a free solute (but still having a fixed conformation), the Gibbs energy changes from $\mu_p^*(\boldsymbol{\phi})$ to $\mu(\boldsymbol{\phi})$. If we want to let the molecule attain *any* conformation, then we must take a proper average over all possible conformations. [This is discussed in more detail in Ben-Naim (2006).]

In theories of protein folding, we usually use the Gibbs energy of the system as a function of ϕ. This function may be referred to as the GEL.

We now list a few properties of the GEL (compare them with the corresponding list of the EL in Subsection 7.1.1):

(1) The function $G(\phi)$ [shorthand for $G(T, P, N; \phi)$ or $G(T, P, N; \phi_1, \ldots, \phi_n)$], defined in Eq. (7.7) is a function of the n variable $\phi_1, \ldots, \phi_n$ when each of the ϕ_i attains the *entire range* of values, $0 \leq \phi_i \leq 2\pi$. This means the GEL is a description or the landscape of the Gibbs energy for *all* conformations, including conformations that are not realizable. This is an important point to remember. For large proteins, the GEL includes all possible knots that one can form with the chain. Although there might be regions on the GEL which are disconnected (i.e. the borders between two such regions involve an infinitely high potential barrier), they are still *parts* of the GEL. One can find in the literature statements like "knots might change the landscape." What one means is that the landscape of a molecule having no knots might be different from the GEL of the molecules having knots. However, for any given protein molecule, the knots *do not change* the landscape — they are parts of the landscape. Another very common statement featuring in the recent literature is that the "landscape theory" has advanced our understanding of protein folding. This is not an exaggeration — it is simply not true. No one knows the shape of the GEL for large proteins. Certainly, there exists no landscape theory of protein folding, and therefore such a "theory" cannot contribute anything to our understanding of protein folding.

(2) Unlike the EL, which to a good approximation, can be assumed to be pairwise-additive, the GEL *does not* have that property. As one can see from Eq. (7.4), the GEL includes the EL as well as a *solvation* Gibbs energy term. The latter is not a pairwise-additive function. It is true that in the literature one can find many expressions of the GEL as if it were pairwise-additive, usually referring to the Kirkwood superposition approximation (KSA) as a justification. Unfortunately, the solvation Gibbs energy $\Delta G^*(\phi)$, and hence the GEL, is not a pairwise-additive function. The assumption of additivity is not a good approximation, it is not a bad approximation, it is not an approximation at all! (See also Appendix L.) This note should be borne in mind whenever one constructs a GEL by assuming pairwise interactions. In particular, this is common to all lattice model of protein. Assuming pairwise additivity for the GEL invalidates automatically the theory which is based on it.

One can split $\Delta G^*(\phi)$ and hence, the GEL into some components or ingredients. The general procedure for doing so is described in detail in Ben-Naim (1989, 1990b, 1992, 2011d), but these ingredients are based on the concept of *conditional* solvation Gibbs energies (see Appendix A), not on pairwise additivity.

(3) Unlike the EL, the GEL depends on the temperature, pressure and composition of the system. In particular, the temperature dependence of the GEL in the definition (7.4) is

$$\Delta S\left(\phi^0 \to \phi\right) = -\frac{\partial \Delta G^*(\phi)}{\partial T} + \frac{\partial \Delta G^*\left(\phi^0\right)}{\partial T}$$
$$= \Delta S^*(\phi) - \Delta S^*\left(\phi^0\right), \qquad (7.11)$$

where $\Delta S^*(\boldsymbol{\phi})$ is the solvation entropy of the protein having a specific conformation, $\boldsymbol{\phi}$. Note that by taking the temperature derivative of ΔG in Eq. (7.4), we assumed that ΔE is temperature independent. The quantity $\Delta S\left(\boldsymbol{\phi}^0 \to \boldsymbol{\phi}\right)$ is the entropy change for the process $\boldsymbol{\phi}^0 \to \boldsymbol{\phi}$ carried out in the solvent at some given T, P, N.

From Eqs. (7.4) and (7.11), we can define the enthalpy change for the same process, i.e.

$$\Delta H\left(\boldsymbol{\phi}^0 \to \boldsymbol{\phi}\right) = T\Delta S\left(\boldsymbol{\phi}^0 \to \boldsymbol{\phi}\right) + \Delta G\left(\boldsymbol{\phi}^0 \to \boldsymbol{\phi}\right)$$
$$= \Delta E\left(\boldsymbol{\phi}^0 \to \boldsymbol{\phi}\right) + \Delta H^*\left(\boldsymbol{\phi}\right) - \Delta H^*(\boldsymbol{\phi}^0).$$
$$(7.12)$$

If we neglect the term $P\Delta V(\boldsymbol{\phi}^0 \to \boldsymbol{\phi})$, where P is the pressure and ΔV is the volume change for the process $\boldsymbol{\phi}^0 \to \boldsymbol{\phi}$, we can define the thermodynamic PEL as

$$\Delta U\left(\boldsymbol{\phi}^0 \to \boldsymbol{\phi}\right) \approx \Delta H\left(\boldsymbol{\phi}^0 \to \boldsymbol{\phi}\right)$$
$$= \Delta E\left(\boldsymbol{\phi}^0 \to \boldsymbol{\phi}\right) + \Delta E^*(\boldsymbol{\phi}) - \Delta E^*\left(\boldsymbol{\phi}^0\right),$$
$$(7.13)$$

where $\Delta E^*(\boldsymbol{\phi})$ is the solvation energy of the solute at conformation $\boldsymbol{\phi}$.

(4) As we have noted in Subsection 7.1.1, the EL is a very complicated function of the variables $\phi_1, \ldots, \phi_n$. Yet, one can confidently say that the GEL is an order-of-magnitude more complicated function. To the best of my knowledge, no one has ever described the GEL function for a protein of, say, 150 amino acids. In my opinion, it is extremely unlikely that such an explicit function will be available in the foreseeable future.

It is also difficult, if not impossible, to even imagine the shape of the GEL.

Yet, surprisingly, people do talk about the *general shape* of the GEL of a protein. (We will discuss this in Section 7.2.) Moreover, people talk about the "GEL theory" and say that it has improved our understanding of the PFP. Such claims are but empty claims. Since there is no "Gibbs energy landscape theory," such a nonexistent theory does not and cannot explain anything!

It should be said that there is nothing wrong with using the GEL in connection with the PFP. However, the focus on the *entire* GEL shape does not help in understanding the PFP. On the contrary, this approach makes a difficult problem vastly more difficult. The reason is that for large proteins the shape of the entire GEL is extremely complicated, and only a tiny part of the GEL is relevant to the process of protein folding in ordinary environments.

7.1.3 The Thermodynamic Potential Energy Landscape

Ignoring for this discussion the difference between the enthalpy change $\Delta H \left(\phi^0 \to \phi \right)$ and the potential energy change $\Delta U(\phi \to \phi)$, we can split the GEL function in two ways:

$$\Delta G \left(\phi^0 \to \phi \right) = \Delta E \left(\phi^0 \to \phi \right) + \Delta G^*(\phi) - \Delta G^* \left(\phi^0 \right),$$
$$(7.14)$$

$$\Delta G \left(\phi^0 \to \phi \right) = \Delta U \left(\phi^0 \to \phi \right) - T \Delta S \left(\phi^0 \to \phi \right).$$
$$(7.15)$$

The first is the relationship between the GEL and the EL based on Figure 7.6. The second is the thermodynamic relationship between the Gibbs energy (or the Helmholtz energy), the enthalpy (or the internal energy) and the entropy change in the specified process. Note that for all of these quantities we exclude the kinetic energies of the solute molecules.

The difference between ΔE and ΔU is important. The former is the change in energy for the process $\phi^0 \to \phi$ carried out in vacuum, or in an ideal gas phase. The latter is the change in energy for the same process carried out in a solvent at some given T, P, N. When the temperature T becomes very small, $T\Delta S$ in Eq. (7.15) may be neglected, and hence

$$\Delta G \xrightarrow{T \to 0} \Delta U. \tag{7.16}$$

However, it is not true that at low temperatures the GEL tends to the EL. Unfortunately, there is great confusion in the literature over the two quantities EL and PEL. It is not uncommon to find statements regarding the low temperature of the GEL being the EL instead of the PEL.

7.1.4 The Potential Energy Surface

In discussing theories of protein folding, one sometimes refers to the EL but means the GEL. This confusion very often arises in connection with the so-called "funnel theory" of protein folding (see Section 7.2).

It should be clear that both the EL and the GEL are well-defined functions of the conformation. The former is strictly relevant to a protein in vacuum. However, since proteins "live" in a solution, and the water is crucial for both the stability and the forces operating on the protein in the folding process, the GEL is the relevant function for protein in a solvent.

In the past, people studied the EL in connection with the search for the global minimum in this function. Once they realized that the relevant function is the GEL rather than the EL, they tried to find an approximation to the GEL. The resulting function is referred to as the PES. The idea is to start with the EL and add "correction" terms to convert it into the GEL. The correct conversion is shown in Eq. (7.14). In order to do this, one needs to know the solvation Gibbs energy of the protein. Unfortunately, these solvation Gibbs energies are unavailable, either from experiment or from any theory.

Assuming (erroneously) that the major part of the solvent-induced contribution to the GEL comes from the various $H\phi O$ effects and assuming (again erroneously) that these $H\phi O$ effects are pairwise-additive, one writes an "approximate" GEL in the form

$$G(\boldsymbol{\phi}) \approx E(\boldsymbol{\phi}) + \sum_{i,j} \delta G_{ij}^{H\phi O}, \qquad (7.17)$$

where $\delta G_{ij}^{H\phi O}$ are the pairwise $H\phi O$ interactions between groups i and j.

This procedure is anything but an approximation to the GEL. As we noted above, the EL is the proper function for studying the conformations of a protein in vacuum. The GEL is the proper function for studying the conformations of a protein in a solution. The so-called PES as defined in (7.17) contains elements of both the EL and the GEL (it is a

Shaatnez of the two), but it is neither the EL nor the GEL. These are two main reasons for this: first, one includes in (7.17) only $H\phi O$ effects ignoring the far more important $H\phi I$ effects (see Chapter 4); second, the $H\phi O$ effects as well as the $H\phi I$ effects are not pairwise-additive functions. (See Appendices L and M.)

In addition, there are more $H\phi O$ and $H\phi I$ effects which are not included in (7.17), such as solvation (of both $H\phi O$ and $H\phi I$ groups) and higher order interactions (between three or more $H\phi O$ and $H\phi I$ groups). The bottom line is that the resulting function referred to as the PES is not an approximation to the GEL, and in fact it is *irrelevant* to the PFP. I am well aware of the fact that some publications report good agreement between experimental data and computed structures based on the PES. In spite of that fact, I still maintain that the PES is an irrelevant function for proteins.

We have defined above four functions: the EL, GEL, PEL and PES. The only relevant function from which one can obtain the *forces* which are exerted on the various groups of the protein is the GEL. Unfortunately, no one knows anything about the explicit form of this function. When people talk about the "landscape theory," one cannot tell whether they are talking about the EL or the GEL. Whatever landscape they are referring to, this is a *function*, not a theory. A recent article by Plotkin and Onuchic (2002), titled "Understanding Protein Folding with Energy Landscape Theory," is in my opinion misleading. One cannot understand "protein folding" with a nonexistent "energy landscape theory."

To the best of my knowledge, there exists neither an EL theory nor a GEL theory. If any reader could send me an outline of such a theory, I would be glad to include it in a future edition of this book.

7.2 A Funnel-like Theory of Protein Folding

As I said at the end of the last section, I am not aware of any *landscape theory* of protein folding. However, it seems to me that when people refer to the landscape theory, they are actually referring to the idea that the landscape (energy or Gibbs energy) has a funnel-like shape.

I do not know the origin of the funnel metaphor used in articles on protein folding. This metaphor has appeared in the title of many recent articles, creating the impression that a new approach or a new theory of protein folding is in progress. Here are some titles:

Wolynes (2001): "Landscapes, Funnels, Glasses and Folding: From Metaphor to Software."
Dill and Chan (1997): "From Levinthal to Pathways to Funnels."
Karplus (2011): "Behind the Folding Funnel Diagram."

The general idea of a funnel-like landscape is probably due to Wolynes *et al.* (1995).

In an article titled "Navigating the Folding Routes," the authors write:

Folded proteins are marvels of molecular engineering. . . . A unique folding pathway, if it exists, could be elucidated with classical chemical experiment. A newer view holds that in the earlier stages a protein possesses a large ensemble of structures. The problem is not to find a single route but to characterize the dynamics of the ensemble through a statistical description of the topography of the free energy landscape. Folding is easy if the landscape resembles a many-dimensional funnel leading through a myriad of pathways to the native structure.

They go on to describe the folding funnel in their Figure 1, saying:

The width of the funnel represents entropy, and the depth, the energy.

They also refer to the "global energy minimum" as the "native structure."

Indeed, "folding is easy if the landscape resembles a many-dimensional funnel. . . ." Unfortunately, no one has shown that for a medium or large protein the landscape has a funnel-like shape. Therefore, the "easy" folding is indeed easy, but in a fictitious landscape.

In some recent articles, one can see many discussions about the "old view" and the "new view" of protein folding, apparently referring to the difference between a unique pathway and multiple pathways to the native structure.

Sometimes the *old view* is associated with a single pathway and the *new view* with multiple pathways. In my view, there is confusion between a *molecular pathway* and *a kinetic pathway* involving intermediate structures. I have discussed these views in Chapter 5.

It is not always clear whether people refer to a folding pathway through a sequence of molecular conformations or a phenomenological description of the kinetics of folding through a sequence of intermediates, each of which could consist of many molecular conformations. The subject of this section is not to discuss the question of a unique pathway

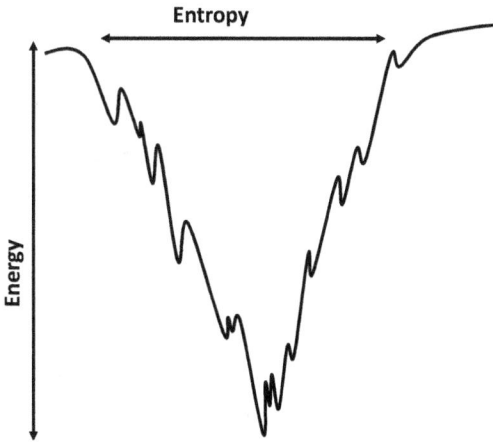

Fig. 7.7. A typical depiction of the funnel model.

versus multiple pathways of folding, but the introduction of the funnel metaphor to describe the folding process. In Figure 7.7, the vertical line represents the energy and the width of the diagram represents the entropy. This is referred to as "the topology of the free energy landscape." Unfortunately, no one has any idea of the "topology" of the GEL. The assumed "many-dimensional funnel" for the GEL is science fiction which has no basis in any statistical theory of proteins.

Karplus (2011) criticizes the funnel diagram by saying:

> To understand the folding process it must be realized that the major determinant of the folding rate is the free energy surface (often referred to as "landscape") of the polypeptide chain, rather than energy shown in the funnel diagram.

As was discussed in Chapter 5, the speed of the folding process is related to the force and the force is related to the gradient of the GEL. It should be stressed that the *speed* of the molecular motion is different from the *rate* of a reaction used in the kinetics of protein folding. It is also important to stress [see also Ben-Naim (2011d)] that the gradient of the GEL does not *determine* the direction of the change of conformation. The gradient determines the *average* force. When this force is very strong, much larger than the random forces exerted by the solvent molecules, we can say that the protein will move, with *high probability* in the direction of the force, i.e. the direction of the gradient in the GEL.

Recognizing that the GEL determines the forces, it is clear that once a protein reaches a minimum in the GEL it will stay there. The length of staying at the minimum depends on the height of the surrounding "wall." Any deviation from the minimum will produce a restoring force toward the minimum. This is exactly the meaning of a stable thermodynamic state. Of course, there are always random forces exerted by the solvent molecules which can kick the protein out of the minimum.

Thus, if the funnel metaphor refers to the shape of the EL, then it is not relevant to the folding in a solvent. If, on the other hand, the funnel describes a local minimum in the GEL, then there is nothing new in the idea of a "funnel" replacing the *minimum*. In fact, the term "funnel" conjures up something through which something is *flown* or "funneled." The minimum in the GEL has by definition a bottom. The funnel does not have a bottom. The bottom here refers to the *local* minimum.

If the funnel metaphor is supposed to describe the entire GEL, then it is not only an inappropriate term for describing a minimum (having a bottom), but it is a totally fictitious description. As we have discussed in Chapter 6, only a small part of the GEL is relevant to the PFP. There is no need to speculate (in fact it is impossible) on the shape of the entire GEL, certainly not to assume that the GEL has a funnel (or any other) shape.

Finally, a comment is in order regarding the funnel diagram shown schematically in Figure 7.7. From this diagram, it is not clear what the "energy" on the ordinate of the figure is. Is this the energy $E(\phi)$ or the PEL $U(\phi)$? In addition, it is not clear what "entropy" shown in the figure means. The entropy of a system is a measure of the width — or, equivalently, the volume of the configurational space — only when all the accessible states are equally probable. This is certainly not true of the protein configurational states. One can still define the Shannon measure of information on the probability of states. In this case, the Shannon measure is not the same as the entropy.

Karplus (2011), who criticized the funnel diagram, also writes:

The free energy is the sum of the potential energy which decreases as the native state is approached and therefore favors folding, and the unfavorable contribution of the decrease in the configurational entropy.

By "the native state is approached" Karplus refers to the gradient in the GEL. That is the main motivation for creating the funnel metaphor to explain the speed of the folding process.

Since the speed is determined (with a higher probability) by the gradient of the GEL, the question arises as to which gradient one takes — that of Eq. (7.14) or Eq. (7.15)? In the first case, taking the gradient of Eq. (7.14) with respect to, say the coordinate of the group at R_1, we split the average force into two parts:

$$-\nabla_1 \Delta G = -\nabla_1 \Delta E - \nabla_1 \Delta G^*(\boldsymbol{\phi}). \qquad (7.18)$$

The first term on the RHS of Eq. (7.18) is the *direct* force exerted on group 1 due to all other groups *within* the protein. The second term is the average solvent-induced force. As discussed in Chapter 5, the latter is probably the larger of the two parts of the average force.

On the other hand, if we take the gradient of Eq. (7.15), we get

$$-\nabla_1 \Delta G = -\nabla_1 \Delta U - (-\nabla_1 T \Delta S). \qquad (7.19)$$

In this case, we split the total average force into parts. One is due to the PEL, i.e. due to the change in the internal energy of the system. The other is due to the entropy change in the system. We know that on the way to the native structure there must be forces on all the groups directing them toward the native structure. However, we do not know the magnitude or the direction of the two parts of the forces in Eq. (7.19). It seems to me that Karplus confuses the two splits of the force in Eqs. (7.18) and (7.19). As we have discussed in Chapter 5, it is likely that the $H\phi I$ forces which are included in the second term on the RHS of Eq. (7.18) are the dominant driving force for the folding process. We do not know even qualitatively the relative magnitude of the two forces in Eq. (7.19). Therefore, the quotation above seems to indicate that Karplus refers to the "potential energy" in Eq. (7.18) "which decreases as the native state is approached," but he adds to this the entropy force in Eq. (7.19) when referring to the "unfavorable contribution of the decrease in the configurational entropy." Thus, in effect, the average force seems to be viewed as two parts: $-\nabla_1 \Delta E$, taken from Eq. (7.18) and $-\nabla_1 T \Delta S$ taken from Eq. (7.19).

To conclude this section, we can say that the GEL function is the relevant function for proteins in solution. This is an extremely complicated function of the angles $\phi_1, \ldots, \phi_n$. For large proteins, there are many regions in the GEL which are inaccessible under normal conditions, and only a tiny part of the GEL is relevant to the PFP. The assumption of a funnel shape like the GEL through which the protein is funneled toward the native state is a fictitious metaphor and has no theoretical basis.

It is unfortunate that "funnels" have become widespread in the literature. In an article by Dill and Chan (1997), a gallery of funnels is shown (some are even given names); see Figure 7.8. These are beautiful figures which deserve to be displayed in an art gallery, but have no place in a scientific article. In the same article, Dill and Chan (1997) refer to one of the landscapes as the Levinthal "golf course"; see [Fig. 7.8(a)]. Regarding the same landscape, Wolynes *et al.* (1995) write:

A flat energy landscape (or golf course) is very unrealistic.

Of course, such a landscape is unrealistic. However, any other funnel-like landscape is as unrealistic as the golf course

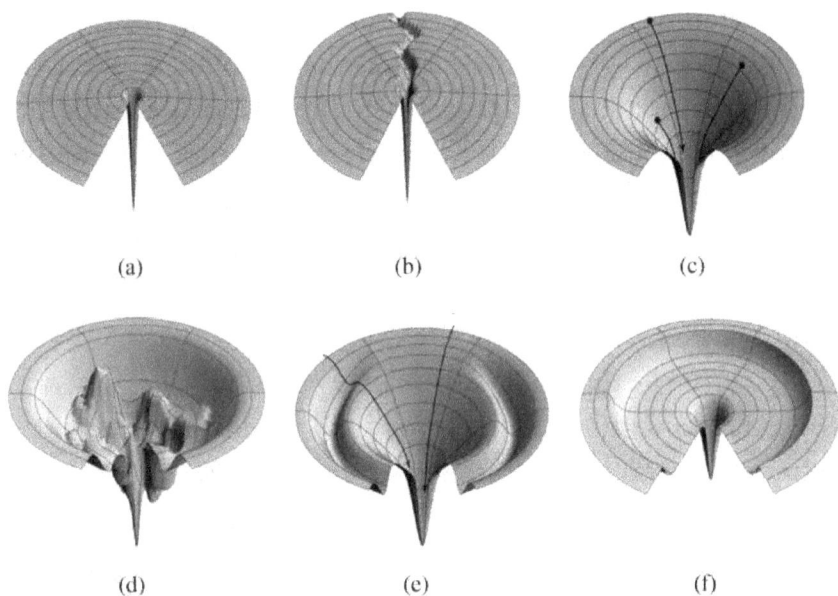

(a)   (b)   (c)

(d)   (e)   (f)

Fig. 7.8. Some funnel-like shapes [reproduced with changes from Dill and Chan (1997)].

landscape. Besides, it is misleading to assign the golf course to Levinthal. In fact, such an assignment is ironic. Levinthal argued that if all the configurations of the protein were visited, then one would reach an absurd result. Specifically, he did not believe that the native structure necessarily resides in the global minimum of the landscape, whether the landscape is portrayed as a golf course or funnel-like. (See Figure 7.8.)

7.3 The Principle of Consistency

The principle of consistency is not a candidate for evolving into a new myth. Yet, I believe that this "principle" fits into this chapter. The reason is the abuse of the term "principle," which in science is almost equivalent to a law, such as the principle of least action, the principle of uncertainty or Le Chatelier's, principle.

The so-called principle of consistency is anything but a *principle*. In an article titled "Theoretical Studies of Protein Folding," Gō (1983) enunciates a new principle: "the principle of consistency of protein folding." His paper begins with the following statement:

> *Most important of all is the fact that the folding process provides a natural mechanism for deciphering information about the three-dimensional structure of a polypeptide chain coded in its amino acid sequence. Implicit in any study of the folding process is the hope that elucidation of its details will ultimately lead to an understanding of the structure of the code.*

This is an extraordinary statement about some (unidentified) "information" encoded in the sequence and that there is a challenge to understand the "structure of the code."

This statement was probably derived from Anfinsen's conclusion known as the thermodynamic hypothesis (Chapter 6), which states that the sequence of amino acid determines the 3D structure of the protein in a *proper environment*, i.e. a given temperature, pressure and solvent composition. Clearly, different environments might lead to different 3D structures and therefore a code, if existing, would depend on the "environment." It follows that the challenge is not only to find the code that translates from sequence to structure, but also to find *infinite possible codes*, if existing, for the infinite possible environments. In addition, it is not clear what is meant by the "*structure* of the code."

Next, Gō defines short-range interactions and long-range interactions:

Here, the short-range interactions mean the intramolecular inter-actions between parts near each other along the peptide chain. Similarly, the term long-range interactions is used to indicate the intramolecular interactions between parts that are separated far from each other along the chain but that are generally close in space.

These definitions of "long-range" and "short-range" are confusing, potentially misleading and superfluous. They are confusing because these terms have already been used in the literature for short- and long-range interactions, depending on the *range*, i.e. the extent of the interactions in terms of distance, respectively. It might also be misleading because the interactions between groups A and B depend on the range (i.e. the distance in space) between A and B and it does not matter where A and B are located along the chain. It also creates the impression that there is some special kind of interaction between A and B according to whether they are near or far *along the chain*. Finally, it is superfluous to introduce these terms. One can simply say the interaction between A and B, which are *n* residues apart, is such and such. The strength of the interaction depends on A and B, and on the *distance* and *orientations* of A and B, and not on *n*.

Following these unfortunate "definitions," one finds an *extraordinary, amusing reasoning* leading to the enunciation of a new principle in protein folding.

First, we have:

The success of methods predicting secondary structures created the impression that short-range interactions alone determine the local conformation of the polypeptide chain (dominance of the short-range interactions). However, it is obvious that the long-range interactions are also important.

Next, the author provides arguments regarding the importance of the long-range interactions and concludes:

The above feeling may be rephrased as an impression of dominance of the long-range interactions, because many pairs of atoms in contact are separated far from each other along the chain.

From these two *impressions* Gō concludes:

The only way to reconcile the apparently conflicting impressions of the dominance of both the short-range and long-range interactions is to assume that these two types of interactions are consistent with each other and therefore individually consistent with the native conformation.

Therefore, the author enunciates:

I propose to call this fact the consistency principle in protein folding structure.

Is this really the only way to reconcile the apparently conflicting impressions? My suggestion is to simply abandon the two *apparent* impressions. This will remove any conflict between the two impressions.

In my view, there exists no conflict between the short- and long-range interactions. There is only an *impression* of a conflict, which is presented as a *fact*. If there exists a real conflict between the two *impressions* on the *dominance* of the short- and long-range interactions, the way to resolve this conflict is to find out which impression is correct and which is incorrect, not to enunciate a new principle.

In my opinion, both *impressions* are incorrect, perhaps even meaningless, and thus there exists no conflict to be resolved. Hence, the enunciated principle is invalid.

The reason for saying so is that no one has created a quantitative measure of the relative importance of one or the other "type" of interaction. (Note that in order to measure the relative importance of the short- and long-range interactions, one must make a clear-cut distinction between the two, which is not made in that article.) Therefore, one cannot claim that one type is more important, and certainly one cannot claim *dominance* of one over the other. What remains, as the author maintains, is only an *impression*, not a *fact*. Gō believes that there exists a conflict between the two impressions, and that "the only way to reconcile the apparently conflicting impressions of the dominance of both the short- and long-range interactions is to assume that these two types of interactions are consistent. . . ."

The existence of long- and short-range interactions in proteins is a fact, as much as the existence of van der Waals, hydrogen bonds, and hydrophobic and hydrophilic interactions. There is no need to introduce a new principle to account for these facts. All of these interactions "live" in perfect harmony (whatever that means) with each other. This is true of proteins, of DNA and of any other molecule.

It is puzzling that this "principle" has been renamed the "principle of minimal frustration," which effectively gives a new meaning to Gō's principle. We will discuss this "new" principle in Section 7.5.

7.4 The Application of Lattice Models for Studying Protein Folding

In the 1930s and 1940s, various lattice models for liquids were very common. Several lattice theories were proposed even for complex liquids such as water. Today, all these models have disappeared. One of the main reasons for the failure of the lattice models for liquids was that they could not account for the so-called communal entropy. Basically, the communal entropy arose because a lattice differs from a liquid in two fundamental aspects. First, particles in a lattice are *localized*; they hold a fixed, or nearly fixed point in the system. In a liquid, each particle can attain all points in the entire volume of the system. The second aspect is associated with the indistinguishability of the particles. Particles in a lattice are distinguishable. Although they are identical, they can be labeled according to the lattice point they occupy. On the other hand, identical particles in a liquid state are indistinguishable; they cannot be labeled. For an informational theoretical discussion on the concept of communal entropy, see Ben-Naim (2008).

It should be said, however, that although the lattice models for liquids are now considered to be obsolete, some lattice models of mixtures are still useful for the study of the *excess properties* of mixtures. The term "excess" here means excess with respect to *symmetrical ideal solutions*. [See Ben-Naim (2006) for more details].

Recently, many lattice models for protein folding have been published. Perhaps the best summary of these approaches may be found in Bryngelson *et al.* (1995):

Simulations of simple protein-like lattice models provide an ideal ground to illustrate the energy landscape ideas. Lattice models have a venerable history. There is widespread agreement that they capture some of the underlying physics of protein folding.

I certainly do not agree that such lattice models can help in any way in understanding the physics of protein folding. In my opinion, all these models will sooner or later be obsolete. This is so not because lattice models are extremely simple compared with proteins, but mainly because not all the types of interactions between the various groups in protein are taken into account. In most of the lattice models, one assumes that the protein is a sequence of two kinds of monomers, say polar and nonpolar, or $H\phi O$ and $H\phi I$ groups, and then one *counts* all the *pairs of contacts* between groups on adjacent lattice points, and assigns pair energies for each type of pair of groups. If you believe that the stability of the structure of proteins and the dynamics of the folding are determined by the solvent, then you cannot represent all these solvent-mediated interactions by *counting* pairs of groups in contact. Secondly, the interactions between two $H\phi I$ groups are not only between pairs but also between triplet and quadruplets of $H\phi I$ groups. Thirdly, even for a specific pair of groups the interaction could vary significantly as a function of their distance, and cannot be represented by a single energy parameter, like pairs of A and B molecules occupying adjacent lattice points. Finally, solvation Gibbs energies are nonpairwise-additive, whereas all lattice models assume pairwise-additive interaction energies.

For all these reasons, I believe that lattice models for protein cannot be used to learn anything about protein. Certainly, simple lattice models *cannot* teach us anything about either the EL or the GEL of proteins. They teach us about the exact form of the EL, for the lattice model, which is irrelevant to proteins in solutions. Though the analysis of lattice models could be a challenging exercise in statistical mechanics, they cannot teach us anything about the main problems of protein folding. It is unfortunate that most of the recent "theories" of protein folding are based on analysis of simple lattice models of protein. They include the "landscape theory," the "funnel theory" and the "principle of minimal frustration" (see next section).

7.5 The Principle of Minimal Frustration

The term "frustration" was borrowed from the theory of spin glasses. According to the dictionary, the word "frustrate" means:

(1) to hinder, prevent or thwart;
(2) to upset, agitate or tire;
(3) to prevent from succeeding;
(4) to balk or defeat in an endeavor;
(5) to induce feelings of discouragement in.

Of course, none of these definitions can be applied — not even figuratively — to a system of particles without further explanation. Consider a system of electrical dipoles

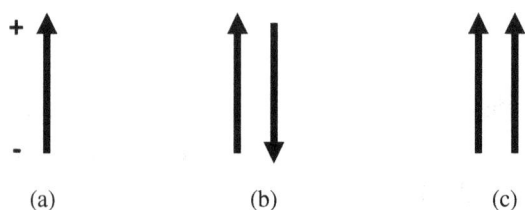

Fig. 7.9. Two possible configurations of two dipoles; (b) antiparallel; and (c) parallel.

(Figure 7.9). Suppose that there are only two possible orientations for the dipole — up or down. When the two dipoles are oriented in opposite directions, the interaction energy is attractive, and when they are oriented in the same direction, the interaction is repulsive. We can figuratively say that the first configuration [Figure 7.9(b)] is favorable and the second [Figure 7.9(c)] is unfavorable. An equivalent way of expressing the same thing is to say that the configuration in [Figure 7.9(b)] is more probable than that in [Figure 7.9(c)].

The simplest example of frustration is shown in Figure 7.10. We have three electrical dipoles (or magnets), such that each dipole can be oriented up or down and that the most favorable (minimal energy) interaction is when the two dipoles are oriented in opposite directions [shown in Figure 7.9(b)].

Now, suppose that we have three dipoles located at the vertices of a perfect triangle (Figure 7.10). In [Figure 7.10(b)], we have all three dipoles upward. In [Figure 7.10(c)] we have two of the dipoles 1 and 2, and the interaction between them is favorable. Now, we look at dipole 3 and we ask: What is the most "favorable" orientation of dipole 3, assuming that the dipoles are at fixed *lattice points* and each can only be in two

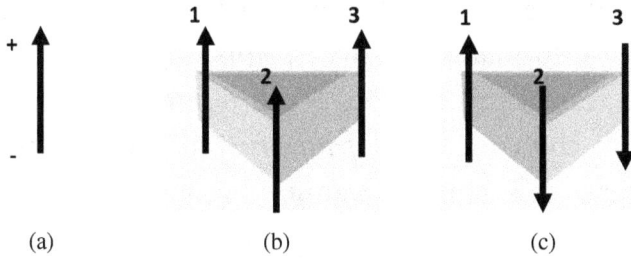

Fig. 7.10. A system of three "frustrated" electric dipoles: (a) the dipole, (b) high energy states, and (c) low energy states.

states of orientation? Clearly, there is no single orientation of dipole 3 which is more favorable.

If dipole 3 is oriented upward, it will interact favorably with dipole 2, but unfavorably with dipole 1. On the other hand, if it is oriented downward [Figure 7.10(c)], it will interact favorably with dipole 1, but unfavorably with dipole 2. Thus, we can say that the system as a whole is *frustrated*; it "wants" to achieve the maximal favorable interaction but it is thwarted from achieving that state because of the constraints imposed on the dipoles (the fixed locations and the only two possible orientations).

Another way of saying — again figuratively — the same thing is that when dipole 3 is oriented upward, it satisfies dipole 2, but frustrates dipole 1. If it is oriented downward, it satisfies dipole 1, but frustrates dipole 2. Thus, whatever dipole 3 "does," it cannot satisfy both dipoles, 1 and 2.

Clearly, in terms of the total potential energy of the three dipoles, there will be two degenerate energy levels. These are the same concept as was used in the theory of neural networks [Amit (1989); Fischer and Hertz (1991)] and in information theory [Matsuda (2000)]. We will further analyze this system in terms of information theory in Appendix N.

In the rest of this section, I will quote a few statements referring to the concept of *frustration* as used for proteins. In my view, this concept cannot be applied to proteins. As is clear from Figure 7.10, in order to have frustration we must have conflicting interactions between particles which are constrained to fixed lattice points. It is not clear what conflicting interactions in protein are, and what the constraints are to a fixed lattice.

In a recent article by Ferreiro, Komives and Wolynes (2013) on "Frustration in Biomolecules," the authors write:

> *Biomolecules are the prime information processing elements of living matter. Most of the inanimate systems are polymers that compute their own structures and dynamics using as input seemingly random character strings of their sequence.*

This opening sentence reminds me of the statement by Lloyd referring to the universe as a computer which computes itself [Lloyd (2006); see also Ben-Naim (2015)]. I do not understand what it means when one says that biomolecules, presumably proteins, "compute their own structures." One can say the same thing about any chemical process, or any process where the system "computes" its dynamics and its final product. These are empty words — biopolymers do not compute anything.

Next, the authors write:

> *In large computational systems with a finite interaction code, the appearance of conflicting goals is inevitable.*

I have no idea what an "interaction code" is, why "conflicting goals" appear, and why their appearance is "inevitable." Since the authors eventually discuss *frustration* in proteins, it is far from clear what the "goal" of protein is.

In the third sentence, the authors reach the concept which is the main topic of the article:

> *Simple conflicting forces can lead to quite complex structures and behaviors, leading to the concept of* frustration *in condensed matter.*

I believe that I understand what "conflicting forces" mean: one force pulling and another pushing can be said to be conflicting forces. But why such conflicting forces lead to "complex structures and behavior" is beyond my comprehension. [Figure 7.11(a)] shows two LJ particles. These can be at favorable and unfavorable configurations. Clearly, in this system, we have conflicting forces: attraction and repulsion. If we force too many of such particles into a volume V, we will find some pairs which attract each other and some pairs which repel each other. Figure 7.12 shows some configurations with "conflicting" interactions. Does this lead to complex structures and behavior?

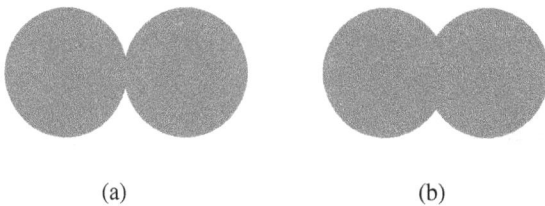

(a)   (b)

Fig. 7.11. (a) Favorable, and (b) unfavorable configurations of two Lennard-Jones spheres.

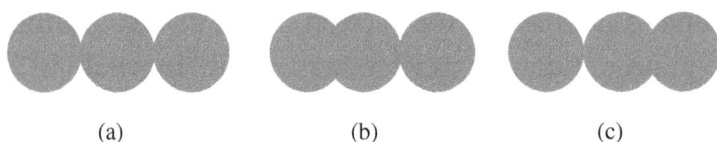

Fig. 7.12. (a) Favorable, (b) unfavorable, and (c) configurations of three Lennard-Jones spheres.

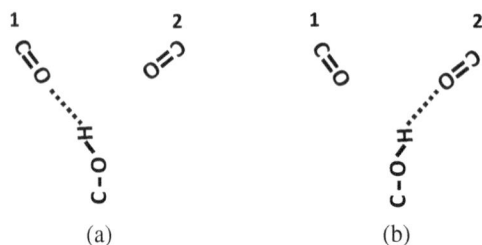

Fig. 7.13. Two possible configurations in the interior of a protein. In (a) a hydroxyl group forms HB with a carbonyl group 1, and (b) it forms a HB with carbonyl group 2.

Figure 7.13 shows two configurations which can occur in protein and to which the term "frustration" can be applied. In [Figure 7.13(a)], a hydroxyl group forms an HB with a carbonyl group 1. We can say that carbonyl group 1 is "satisfied," whereas carbonyl 2 is "frustrated." On the other hand, a simple rotation about the C–O bond of the hydroxyl group can lead to satisfaction of carbonyl 2, while frustrating carbonyl 1. Does this lead to complex structures and behavior?

Furthermore, the authors relate the concept of frustration to the landscape theory:

The energy landscape theory of protein folding provides a framework for quantifying frustration. . . .

Since, to the best of my knowledge, there exists no landscape theory of protein folding, I cannot see how such theory, even if it existed, could provide quantification of the concept of frustration.

The final sentence of the abstract states:

> *We hope to illustrate how* frustration *is a fundamental concept in relating function to structural biology.*

This sentence inspired me to read carefully the 100-page article in its entirety. I did not find *any explanation* as to how the concept of frustration as used in spin glass theory can be implemented in protein, and I was not convinced that this is a fundamental concept. In addition, I did not see how this concept relates "function to structural biology."

The notion of a landscape theory of protein folding, which appears many times throughout the article, is explained nowhere. In some places, the authors refer to the lattice models of protein, which in my view are models irrelevant to protein (see Section 7.4).

Another extraordinary statement relating "frustration" to "protein structure prediction" is the following:

> *Understanding frustration globally for real proteins provides a "license to do bioinformatics," enabling us to extract physical information about molecular forces from the structural database. Studying frustration thus provides practical tools for protein structure prediction. . . .*

All these statements and many others left me deeply *frustrated*. Reading the entire article, I could not find a simple *definition* or *explanation* of the term "frustration." I also failed to see how *frustration* helps in any way in *prediction* of protein structures.

Here is another vague statement, in the article by Ferreiro *et al.* (2013):

> *The third Law of Thermodynamics, which required entropy to vanish at absolute zero, implying a unique ground state.*

This is not true. There are many systems which have a residual entropy, which means multitudes of states at $T = 0$. Examples are a solid of NO, CH_3D, water and many others. Here there are no conflicting interactions, no frustration. The residual entropy is simply a result of many configurations having nearly the same energy.

One can say that frustration, as defined in spin glasses, leads to multiple degeneracy of the ground state. The inverse of this statement is not necessarily true. Systems having residual entropy, such as NO or water, have multiple degeneracy, but these systems cannot be said to be frustrated.

> *Since Kauzmann, the hydrophobicity of side chains has been known to be the dominant effect in determining protein architecture.*

This is "well known" indeed, but unfortunately it is not well known that this statement is false. The authors ignore the fact that more than 20 years ago several articles were published showing that hydrophobic effects are not dominant, and the $H\phi I$ effects are likely to be much more important in determining both the structure stability and the dynamics of the folding process. (See Chapter 4.)

A little frustration can be good, a lot can be disastrous.

I have no idea what this sentence means. In my experience, reading the article on "frustration in biomolecules" was highly *frustrating*, but not disastrous. This is, of course, my personal experience. What this quoted sentence means in the context of protein is not clear

> *Frustration due to topology alone is thus only a small part of the story of the difficulty of finding the global free energy minimum for protein.*

Why should one search for the global free energy minimum (see Chapter 6) in the first place, and why this is related to frustration? In my view, there is nothing to connect the concept of frustration to the global free energy minimum.

I also fail to understand the concept of "global and local frustration."

In the entire manuscript, the authors refer to the original paper by Bryngelson and Wolynes (1987) for a definition of the "principle of minimal frustration." Unfortunately, this paper does not contain a definition of frustration. The authors admit that they borrowed the concept of frustration from spin glass theory. However, nowhere in the article do they explain how this concept is implemented in proteins, how it is quantified and how they arrived at the "principle of minimal frustration."

Most surprisingly (and to me also frustratingly), the authors relate this principle to Gō's principle of consistency, which we discussed in Section 7.3.

In Bryngelson and Wolynes (1987), the authors explicitly say:

This feature was used by Gō in his models of folding [Gō (1983)] and we make use of it also. We call it the principle of minimal frustration.

In a review article, Shakhanovich (2006) writes that Bryngelson and Wolynes *adopted* the "consistency principle," proposed by Gō (1983, 1985), and refers to it as the "principle of minimal frustration." He then explains the Bryngelson and Wolynes theory: ". . . each amino acid can be either in its native conformation or in any of v non-native ones. . . assuming that when amino acids are in their native conformations their intrinsic energy. . . is lower than for interacting amino acids that adopt non-native conformation." Shakhnovich also makes some critical comments on Gō's "maximal consistency," saying, "Such a postulate is not entirely physical."

Unfortunately, Shakhanovich and Gutin (1989) proposed a model of a "microscopic Hamiltonian" which is constructed on the assumption of pairwise interactions. In my view, such a Hamiltonian is also "far from being entirely physical."

Thus, in the entire article by Ferreiro *et al.* (2013), I could not find any definition of "frustration" which is suitably applied to protein. I do not believe that this concept as used in spin glasses can be applied to protein. Secondly, I do not think that this concept can be quantified. [There were suggestions by Frerrero *et al.* (2007) to define a "frustration index," but I failed to see how this index would have anything to do with frustration as defined in spin glasses.] Finally, I failed to see a clear-cut explanation for the "principle of minimal frustration." If it is true, as the authors admit, that the principle of minimal frustration is the same as the principle of consistency, then my conclusion in Section 7.3 applies to the principle of minimal frustration.

In many articles, the principle of minimal frustration is quoted as a well-established principle in the theory of protein folding. In Levy *et al.* (2005), we find:

The funnel energy landscape, a consequence of the minimal frustration achieved by evolution in sequences, explains how most proteins fold efficiently and robustly to their functional structure. . . .

The "funnel energy landscape" is a false concept which has nothing to do with frustration and certainly was not "achieved by evolution."

In Ferreiro *et al.* (2011), we find:

Overall, strong energetic conflicts are minimized in native states satisfying the principle of minimal frustration. Local violations of this principle open up possibilities to form complex multifunctional energy landscapes.

Prentiss *et al.* (2006) write:

The principle of minimal frustration states that native contacts must be more favorable, in a strict statistical sense, than non-native contacts in order for proteins to fold on physiological time scales.

The claim that native contacts must be more favorable than non-native contacts is trivial. It is not clear, however, what this statement has to do with frustration as defined in spin glasses. In addition, it is not clear what it has to do with Gō's principle of consistency. It is far from clear how the favorable native contacts affect the timescale of the folding process. In Yang *et al.* (2004), we find:

> *It is now well established that proteins are minimally frustrated. The minimal frustration principle states that naturally occurring proteins have been evolutionarily designed to have sequences that achieve efficient folding to a structurally organized ensemble of structures with few traps arising from discordant energetic signals.*

As noted earlier the principle of minimal frustration is far from being well defined, let alone well established. The above-quoted statement involves concepts such as "evolution-arily designed," "efficient folding" and "discordant energetic signals," which only add to the fuzziness of the concept of frustration when applied to proteins. This statement clearly has nothing to do with the frustration concept as used in spin glass theory.

In Jenik *et al.* (2012), we find a new description of the principle of minimal frustration:

> *This principle states that the general energy of the protein decreases more than what may be expected by chance as the protein assumes conformations progressively more like the ground (native) state.*

This is an extremely vague sentence. I have no idea what "the general energy of the protein" is. What is the decrease in the general energy "expected by chance"? Why is the ground state the native state? (See Chapter 6.) Finally, what have all these statements or descriptions of the principle of minimal frustration got to do with frustration as defined in spin glasses?

Another "explanation" of the principle is found in Wolynes *et al.* (1995):

That the interactions of a kinetically foldable protein must have fewer conflicts than typically expected is known as the principle of minimal frustration.

In an article by Bryngelson *et al.* (1995), we find yet another new description of the "Bryngelson and Wolynes. Bryngelson and Wolynes used the term "the principle of minimal frustration" in describing the smoothness postulate, insofar as it is what distinguished natural proteins from random heteropolymers. The smoothness of folding landscapes arises from the selection of protein sequences by evolution.

This is again a new description of the principle which is unclear in itself. It is also unclear what it has to do with the concept of frustration, as originally borrowed from the field of spin glasses.

At the end of the same paragraph, the authors write:

Consistency between secondary structures and global tertiary structures is also important. This is the "principle of structural consistency" enunciated by Gō.

In an article by Onuchic *et al.* (2004), we find:

Evolution achieves robustness by selecting for sequences in which the interactions present in the functionally useful structure are not in conflict, as in a random heteropolymer, but instead are mutually supportive and cooperatively lead to a low-energy structure. The interactions are "minimally frustrated" or "consistent."

This is again a new formulation of the principle of minimal frustration, which is based on an ill-defined "conflict" between interactions, "supportive" and "cooperative" interactions, and besides has nothing to do with the concept of frustration as defined in spin glasses.

258 | Myths and Verities in Protein Folding Theories

In Wikipedia on "protein folding," we find:

The principle of minimal frustration states that nature has chosen amino acid sequences so that the folded state of the protein is very stable.

This is trivial. Everyone knows that proteins are stable! But what has it got to do with frustration? Clearly, all these descriptions do not consist of a well-defined concept of frustration and certainly do not provide any explanation for the principle of minimal frustration.

I have spent a great deal of time trying to understand how frustration as defined in spin glasses is adapted to proteins. I wrote to the authors who write on the principle of minimal frustration. I spoke with people who work on protein folding, but no one could explain to me what this principle means. Therefore, I concluded that a better, more appropriate principle in this field is the "principle of *maximum* frustration," where "frustration" means what everyone knows, as described at the beginning of this section.

In most cases, the authors refer to the original works of Bryngelson and Wolynes (1987, 1989). Unfortunately, nowhere in these articles can one find how the well-defined *frustration* in spin glasses is adapted to proteins. In my opinion, the term "frustration" as defined for spin glasses cannot be used for proteins. The interactions between various groups of the proteins do not resemble the case of dipoles situated on lattice points. The fact that there are so many conflicting descriptions of the same principle is a vivid manifestation of the confusion in the entire field of theories of protein folding based on landscapes, funnels and frustrations.

All these elements of confusion appear in a recent article by Noel *et al.* (2010):

> *In a minimally frustrated, funnel-like energy landscape, one expects that native contacts are on average favorable and dominate over non-favorable non-native ones. Topological constraints imposed by the existence of native knots radically alter the funneled landscape.*

First, as pointed out earlier, the *energy landscape* is irrelevant to proteins in solutions. Second, a "funnel-like energy landscape" is a fictitious concept which at best means a minimum in the GEL. Third, I have no idea what "a minimally frustrated funnel-like energy landscape" means. Fourth, if knots exist in the native structure of the protein, they, as well as other possible knots are *part* of the GEL; they do not *alter* the landscape.

Quite recently, in an article by Morcos *et al.* (2014), we have seen yet another new "explanation" for the principle:

> *The key to our analysis is the principle of minimal frustration (3,5) which states that, for quick and robust folding, the energy landscape of a protein must be dominated by interactions found in the native conformation.*

This statement is supposed to *explain* what the principle of minima frustration is. Instead, it is a statement on the EL being dominated by interactions found in the native conformations. In my opinion, such statements are not only not true, but also add to the confusion already existing in connection with the concept of frustration in protein folding. It is not clear why the EL should be dominated by interactions found in the native conformation. The EL and the GEL are determined by *all interactions in all configurations* of the protein. The native conformation occupies a tiny region in the EL. In addition, it is not clear what the interactions found in the native conformation have got to do with the "quick and robust" folding. In fact, I believe that the quick and robust folding depends on the interactions found in all the conformations leading from the unfolded to the folded state. Finally, I fail to see how all these descriptions of the principle of minimal frustration have anything to do with frustration as defined in spin glasses. I should also add that whatever the merits of this description of the principle are, it is certainly different and perhaps inconsistent with other descriptions given by various authors, some of them quoted above.

To conclude, the concept of frustration was originally introduced for dipoles having two states and being constrained to specific lattice points. Groups on proteins are not dipoles and they are not constrained to some fixed lattice.

Thus, unless the term "frustration" is clearly defined and quantified in terms of molecular interactions between groups of the protein, I do not see any reason to use this term, and certainly not the new principle in connection with proteins. As I have written in other places in this book, I would welcome any clear and convincing explanation of how "frustration" is defined for proteins, how it is quantified, and what the principle of minimal frustration is.

7.6 The Structure–Function Paradigm

I would like to conclude this chapter with a short note on the long-standing paradigm which basically states that, in order to function, proteins must have a well-defined 3D structure. This relationship is sometimes referred to as the structure–function paradigm.

Of course, there is nothing wrong with this paradigm as long as one refers to those proteins which do have structure and their structure is crucial to their activity. However, recently, more and more proteins have been discovered which have very important functions [for a review, see Uversky and Dunker (2010)] but which either have several different stable structures or have a considerable part of their chain with no well-defined structure. These proteins are referred to as IDPs.

IDPs have become one of the most interesting fields of research on proteins. This field does not replace the existing paradigm, but rather causes the structure–function paradigm to be more restricted, or to have less general validity.

The discovery of IDPs actually defies the structure–function dogma [Tompa and Han (2012)]. In fact, it also adds to the argument discussed in Chapter 6, that there is no *folding code.*

> Some IDPs derive their functionality directly from structural disorder and are described as having entropic chain functionality.

Here, the authors use the term "entropic" as being synonymous with "disordered."

The discovery of IDPs also has some effects on other aspects of protein folding. First, it removes some of the pressure and perhaps also the urgency of solving the PFP. Clearly, it considerably reduces the probability that a folding code exists and, furthermore, from what I have gathered from some review articles, it seems that IDPs contain more

$H\phi I$ side chains than in structured proteins. Perhaps this fact will encourage people to pay more attention to various $H\phi I$ effects, which, as we have seen in Chapters 4 and 5, are far more important to both the stability and the dynamics of proteins.

Finally, I would like to add the following quotation from Tampa and Han (2012):

> *Once upon a time, it was a dogma that proteins function only if they acquire their natural, or native, structure.... The immense success and explanatory power of the structure–function paradigm is witnessed by the more than 80,000 protein structures in the Protein Data Bank, and by countless works that have elucidated the function of enzymes, receptors and the so-called structural proteins.... Nonetheless, we now understand that for many important proteins ... the native, functional state is unstructured. Those intrinsically disordered proteins (IDPs) defy the structure–function model.*

I fully agree! I would add to this that IDPs also debunk the myth of the existence of a folding code.

Appendix A

The Definition of the Solvation Process and the Corresponding Thermodynamic Quantities

Traditionally, solvation was studied within the thermodynamics of solutions. There were essentially three competing definitions of the solvation processes, and the corresponding thermodynamic quantities. All these pertain to very dilute solutions of, say, a solute A in a solvent B.

Within thermodynamics it was impossible to argue about the relative significance of the three definitions. During the 1970s, I showed that one of these definitions of the solvation thermodynamic quantities has a simple interpretation in terms of statistical thermodynamics. This was the one based on the molarity concentration (see below). It was argued that this "molarity scale" has an advantage as a measure of the solvation thermodynamics.

A few years later, the very definition of the *solvation process* was changed. The definition led to a generalization of the concept of solvation and to the application of solvation thermodynamics to a far greater range of systems.

We will briefly compare one of the older definitions of the solvation process with the new definition. [For more details, the reader is referred to Ben-Naim (2006).]

A.1 The Old Definition of the Solvation Process

Consider a very dilute solution of A in a solvent B. Assuming also that the vapor above the solution is an ideal gas mixture, we can write the chemical potential of A in the two phases as

$$\mu_A^l = \mu_A^{0l} + RT\ln\rho_A^l, \tag{A.1}$$

$$\mu_A^g = \mu_A^{0g} + RT\ln\rho_A^g. \tag{A.2}$$

Here, ρ_A^l and ρ_A^g are the *molar* concentrations of A in the liquid and the gaseous phase, respectively. μ_A^{0l} and μ_A^{0g} are referred to as the *standard* chemical potential of A in the two phases. In the old literature, the latter quantities were interpreted as the chemical potential of A in a solution of one molar concentration. This interpretation is not warranted, for the following reason. Equations (A.1) and (A.2) are valid only for very dilute solutions of A and B (in the liquid phase). At one molar concentration, it is not clear whether Eq. (A.1) is valid. Therefore, the interpretation of μ_A^{0l} and μ_A^{0g} remains vague.

Consider the transfer of one mole of A from the gaseous phase to the liquid phase. The Gibbs energy change for this process is

$$\Delta G(\rho - \text{process}) = \mu_A^l - \mu_A^g$$

$$= \mu_A^{0l} - \mu_A^{0g} + RT \ln \left[\frac{\rho_A^l}{\rho_A^g} \right]. \quad (A.3)$$

For the particular case where $\rho_A^l = \rho_A^g$, we get the standard *solvation Gibbs energy based on the molar concentration scale*:

$$\Delta G_A^0(\rho - \text{process}) = \mu_A^{0l} - \mu_A^{0g}. \quad (A.4)$$

Note that this standard quantity applies only for very dilute solutions of A in B.

The molecular interpretation of ΔG_A^0 is simple. It is the Gibbs energy change for transferring one A molecule from a fixed position in the gaseous phase to a fixed position in the liquid phase at constants T and P, and where the solute A is very diluted in the solvent B. We have defined the process for one A molecule. By multiplying the corresponding Gibbs energy change by the Avogadro number, we get the thermodynamic quantity as defined in Eq. (A.4) for one mole of A.

A.2 The New Definition of the Solvation Process

We now use the molecular interpretation of ΔG_A^0 to *define* a new process of solvation. This is the process of transferring one A molecule from a fixed position in the gaseous phase to a fixed position in the liquid phase [Figure A.1(a)]. The corresponding Gibbs energy change may be obtained from

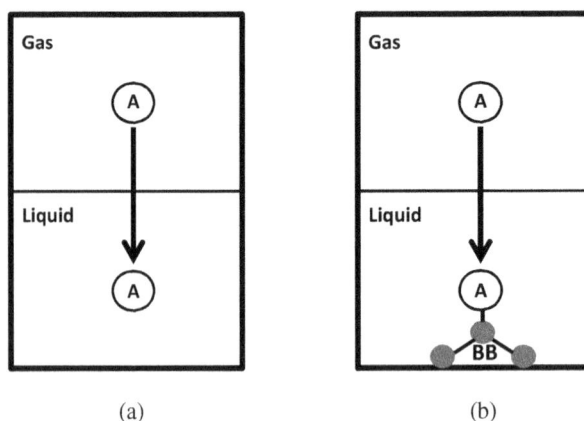

Fig. A.1. (a) Solvation and (b) conditional solvation of a solute A next to a backbone (BB).

the general expression for the chemical potential, in any phase:

$$\mu_A = \mu_A^* + k_B T \ln \rho_A \Lambda_A^3. \tag{A.5}$$

Here, ρ_A is the *number* density, $\rho_A = N_A/V$, Λ_A^3 is the momentum partition function, k_B is the Boltzmann constant, and μ_A^* is the pseudochemical potential. The last is defined as the Gibbs energy change for the process of inserting one A molecule into a fixed position in the liquid. If we are interested in the chemical potential per mole, we simply multiply by the Avogadro number. We now define the Gibbs energy change for the transfer of one A molecule from a fixed position in the gaseous phase to a fixed position in the liquid phase by

$$\Delta G_A^* = \mu_A^{*l} - \mu_A^{*g}. \tag{A.6}$$

Note that this quantity is defined for any concentration of A in the two phases. In the limit of dilute ideal solutions, ΔG_A^* as defined in Eq. (A.6), is identical with ΔG_A^0 as defined in Eq. (A.4). However, it should be noted that these two quantities pertain to two different processes of solvation. In addition, the quantity ΔG_A^* is defined for any concentration of A in B including pure A.

There are other important differences between the traditional solvation quantities and the new one. A detailed discussion may be found in Ben-Naim (1992, 2006). Here, we discuss only one important difference.

When we take the temperature dependence of ΔG_A^* as defined in Eq. (A.6), we get

$$\Delta S_A^* = -\frac{\partial \Delta G_A^*}{\partial T}. \qquad (A.7)$$

ΔS_A^* is the entropy change for the solvation process as defined for ΔG_A^*, i.e. the entropy change for transferring one A molecule from a fixed position in the gaseous phase to a fixed position in the liquid phase. On the other hand, taking the temperature derivative of ΔG_A^0, defined in Eq. (A.4), does not in general give the entropy change for the same process for which ΔG_A^0 is defined.

The definition of the solvation process and the corresponding thermodynamic quantities can be easily extended to the conditional solvation. The latter process is the relevant one for protein folding (see Chapter 2). The conditional solvation is defined as the process of transferring a solute A from a fixed position in an ideal gas to a fixed position near some backbone (BB) in the liquid phase [Figure A.1(b)].

The formal definition of the solvation thermodynamic quantities is discussed in Chapter 3. Similarly, the conditional solvation quantities are defined in a similar way except for the presence of the BB.

Before we discuss the solvation quantities for molecules with internal rotational degrees of freedom, I would like to add a comment on another definition of the solvation thermodynamic quantities.

As I mentioned above, before the 1970s, most solution chemists used various *standard* solvation quantities based on different choices of concentration scales. No one used *excess thermodynamic quantities* for solvation. Excess quantities were reserved to describe the excess with respect to various ideal solutions (ideal gas, ideal dilute solutions and symmetrical ideal solutions). [See Ben-Naim (2006) for details.]

Soon after the new definition of the solvation process was published [see Ben-Naim (1987)], most people recognized the advantages of this approach, and the so-called Ben-Naim standard process of solvation was widely accepted and used.

A few scientists noticed that some excess quantities have a similar expression to the solvation Gibbs energy as defined above. They took the new definition of solvation, as described above, and referred to it as the excess chemical potential, while ignoring the existing new definition. The difference between the solvation quantities and the excess quantities is discussed in great details in Ben-Naim (2006, 2009). Here, I will make only a few comments about the use of excess functions for solvation quantities.

First, the excess quantities do not correspond to a well-defined *process*. There are different excess quantities defined with respect to different ideal systems. Second, the excess chemical potential (with respect to ideal gas) is identical with the solvation Gibbs energy. However, while all the derivatives of the *solvation* Gibbs energy provide other thermodynamic quantities of solvation (such as ΔS_s^*, ΔH_s^* or ΔV_s^*), for the same process, it is far from clear how to obtain these solvation quantities from the excess functions. Third, as noted in Chapter 5, the definition of the hydrophobic as well as hydrophilic interactions is related to the process of solvation of monomers and "dimers." It is not straightforward to define hydrophobic interactions with the use of excess functions. Finally, the important quantity of conditional solvation is a natural extension of the solvation process, and it is not clear how to define conditional solvation using excess functions.

For these, as well as other reasons, I suggest to use the solvation process for the solvation thermodynamics, and to reserve the excess function for describing *deviations from ideal solutions* [see Ben-Naim (2006)].

A.3 Solvation Gibbs Energy of a Molecule with Internal Rotations

In the previous sections, we discussed the solvation thermodynamics of a molecule having no internal degrees of freedom, or, if it has, the internal partition function does not depend on the presence of the solvent.

In this case, we have

$$\mu_S = \mu_S^* + RT \ln \rho_S \Lambda_S^3 \tag{A.8}$$

and the solvation Gibbs energy is given by (see Chapter 2)

$$\Delta \mu_S^* = \mu_S^{*l} - \mu_S^{*g} = -k_B T \ln \langle \exp (\beta B_S) \rangle_0. \tag{A.9}$$

Most solutes of interest have internal degrees of freedom and these are affected by the presence of the solvent. We will

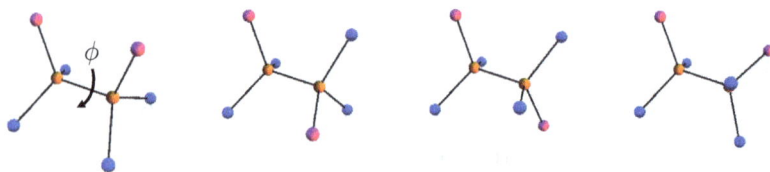

Fig. A.2. A finite number of conformations of substituted ethane.

treat here a simple example; a molecule with one internal rotation ϕ. For simplicity, we assume that ϕ can attain only a finite number of angles ϕ_i, with $i = 1, 2, \ldots, n$, (Figure A.2). Also for simplicity, we assume that the internal energy levels are the same for each of these isomers but the rotational partition function is different for each isomer.

We are interested in the expression for the solvation Gibbs energy of the solute s.

It is not uncommon to find in the literature an equation of the form

$$\Delta G_s^0 = \langle \Delta G^0(i) \rangle, \qquad (A.10)$$

where ΔG^0 is referred to as the solvation Gibbs energy of the ith conformer, and ΔG^0 is the solvation Gibbs energy of the solute s. An equation such as (A.10) is meaningless unless one first specifies which process of solvation is referred to (i.e. which of the quantities based on the choice of standard states, or whether it is the solvation process described in Section A.1, and which distribution one uses to define the average quantity. In this section, and in the rest of the book, we will always refer to the process of solvation as that defined in Section A.1, and therefore $\Delta G^*(i)$ or $\Delta \mu^*(i)$ is the change in Gibbs energy for the transfer of one s molecule at the specific conformation ϕ_i

from a fixed position in an ideal gas phase to a fixed position in a liquid.

The next question referring to Eq. (A.10) is how one takes the average, i.e. which probability distribution is used in the definition of the average on the RHS of (A.10). Once one specifies what one means by $\Delta G^0(i)$, and how one takes the average, the RHS of (A.10) is a well-defined *average* of all $\Delta G^0(i)$. The main question still left open is: Is this average equal to the *experimental* Gibbs energy of solvation of the solute s?

We now derive the relationship between the *experimental* solvation Gibbs energy of the solute s, and the solvation Gibbs energies of all the conformers.

For each conformer, we write its chemical potential in the liquid mixture as

$$\mu_i^l = \mu_i^{*l} + k_B T \ln \rho_i^l \Lambda_i^3, \qquad \text{(A.11)}$$

where

$$\mu_i^{*l} = U(i) - k_B T \ln q_i - kT \ln \langle \exp[-\beta B(i)] \rangle. \quad \text{(A.12)}$$

$U(i)$ is the internal rotational potential of the conformer i measured with respect to some chosen reference conformation, q_i is essentially the rotational PF of i, and $B(i)$ is the binding energy of i.

At equilibrium we must have the equality

$$\mu_s = \mu_1 = \mu_2 = \cdots \mu_n. \qquad \text{(A.13)}$$

(See also Appendix K.)

This equation means that the change in the Gibbs energy of the system when we add one solute s is the same as adding one specific conformer i. Since the system is at equilibrium, the chemical potential μ_i does not depend on the index i.

Next, we write for the solute s

$$\mu_s^l = \mu_s^{*l} + k_B T \ln \rho_s^l \Lambda_s^3, \qquad (A.14)$$

where $\rho_s^l = \sum \rho_i$. Note that for each i, the momentum partition function Λ_i^3 is equal to Λ_s^3. From the equilibrium condition (A.13) and Eqs. (A.12) and (A.14), we have the equality for each i

$$\mu_s^{*l} + k_B T \ln \rho_s^l \Lambda_s^3 = \mu_i^{*l} + k_B T \ln \rho_i^l \Lambda_i^3, \qquad (A.15)$$

or, equivalently,

$$\mu_s^{*l} - \mu_i^{*l} = k_B T \ln x_i^{eq}, \qquad (A.16)$$

where $x_i^{eq} = \rho_i^l/\rho_s^l = N_i^l/N_s^l$ at equilibrium. Note that x_i^{eq} is defined as the mole fractions of *solutes* in conformation i (i.e. N_i^l divided by N_s^l, not by the total number of molecules in the system). x_i^{eq} is the equilibrium mole fraction of the conformer i. Since $x_i^{eq} \leq 1$, we always have

$$\mu_s^{*l} - \mu_i^{*l} < 0. \qquad (A.17)$$

From Eq. (A.16), we can get an equation relating μ_s^{*l} to all the μ_i^{*l}. We write (A.16) as

$$x_i^{eq} = \exp\left[\beta\left(\mu_s^{*l} - \mu_i^{*l}\right)\right]. \qquad (A.18)$$

Summing over i and noting that

$$\sum x_i^{eq} = 1, \qquad (A.19)$$

we get

$$\exp\left[-\beta\mu_s^{*l}\right] = \sum_{i=1}^{n} \exp\left(-\beta\mu_i^{*l}\right). \qquad (A.20)$$

This is an important relationship. It relates μ_s^{*l} to all the individual pseudochemical potentials, μ_i^{*l}.

We can write a similar equation for the gaseous phase. We will use the notation y_i^{eq} for the equilibrium distribution of the conformers in the ideal gas phase. As in Eq. (A.18), we have

$$y_i^{eq} = \exp\left[\beta\left(\mu_s^{*g} - \mu_i^{*g}\right)\right], \qquad (A.21)$$

and summing over i, we get the analog of Eq. (A.20):

$$\exp\left(-\beta\mu_s^{*g}\right) = \sum_{i=1}^{n} \exp\left(-\beta\mu_i^{*g}\right). \qquad (A.22)$$

Taking the ratio of Eqs. (A.20) and (A.22), we get

$$\exp\left[-\beta\left(\mu_s^{*l} - \mu_s^{*g}\right)\right] = \frac{\sum \exp\left(-\beta\mu_i^{*l}\right)}{\sum \exp\left(-\beta\mu_i^{*g}\right)}. \qquad (A.23)$$

Since

$$y_i^{eq} = \frac{\exp\left(-\beta\mu_i^{*g}\right)}{\sum \exp\left(-\beta\mu_i^{*g}\right)}, \qquad (A.24)$$

we can write Eq. (A.23) as

$$\exp\left(-\beta\Delta\mu_s^*\right) = \sum_{i=1}^{n} y_i^{eq} \exp\left(-\beta\Delta\mu_i^*\right). \qquad (A.25)$$

This is the required relationship. On the LHS of Eq. (A.25), we have the solvation Gibbs energy of the solute s, $(\Delta\mu_s^*)$. On the RHS of Eq. (A.25), we have the individual solvation Gibbs energies of the conformers $(\Delta\mu_i^*)$.

Note carefully that $(\Delta\mu_s^*)$ is not a simple average of all the $\Delta\mu_i^*$, but

$$\exp(-\beta\Delta\mu_s^*) = \langle\exp(-\beta\Delta\mu_i^*)\rangle. \qquad (A.26)$$

The average in Eq. (A.26) is taken with the distributions given in Eq. (A.24).

Another relationship may be obtained by dividing Eq. (A.18) by Eq. (A.21),

$$\frac{x_i^{eq}}{y_i^{eq}} = \exp[\beta(\Delta\mu_s^* - \Delta\mu_i^*)], \qquad (A.27)$$

or, equivalently,

$$y_i^{eq} = x_i^{eq}\exp(-\beta\Delta\mu_s^*)\exp(+\beta\Delta\mu_i^*). \qquad (A.28)$$

Summing over i and rearranging, we get a relationship equivalent to Eq. (A.25), but with the distribution x_i^{eq}, i.e.

$$\exp(+\beta\Delta\mu_s^*) = \sum_{i=1}^{n} x_i^{eq}\exp(+\beta\Delta\mu_i^*)$$
$$= \langle\exp(\beta\Delta\mu_i^*)\rangle_x. \qquad (A.29)$$

Note the different averages in Eqs. (A.26) and (A.29)

Note also that when $\beta \Delta \mu_s^* \ll 1$ and all $\beta \Delta \mu_i^* \ll 1$, we can get from Eq. (A.25) the approximation

$$1 - \beta \Delta \mu_s^* \approx \sum y_i^{eq} \left(1 - \beta \Delta \mu_i^*\right) \tag{A.30}$$

or, equivalently,

$$\Delta \mu_s^* \approx \sum_i y_i^{eq} \Delta \mu_i^*. \tag{A.31}$$

In this approximation, we can also write from Eq. (A.28)

$$y_i^{eq} \approx x_i^{eq} \left(1 - \beta \Delta \mu_s^* + \beta \Delta \mu_i^*\right). \tag{A.32}$$

Hence, instead of (A.31), we also have

$$\Delta \mu_s^* \approx \sum_i x_i^{eq} \Delta \mu_i^*. \tag{A.33}$$

Thus, only for very small $\beta \Delta \mu_s^*$ and $\beta \Delta \mu_i^*$ is the solvation Gibbs energy of s, a simple average of the solvation Gibbs energies of all the conformers.

A.4 The Solvation Volume

We start with Eq. (A.20) and take the pressure derivative to get

$$\exp\left(-\beta \mu_s^{*l}\right) \left(\beta V_s^{*l}\right) = \sum_i \exp\left(-\beta \mu_i^{*l}\right) \left(\beta V_i^{*l}\right). \tag{A.34}$$

Using Eq. (A.18), we rewrite Eq. (A.34) as

$$V_s^{*l} = \sum x_i^{eq} V_i^{*l}. \tag{A.35}$$

Similarly, for the gaseous phase, we have [from Eqs. (A.21) and (A.22)]

$$V_s^{*g} = \sum_i y_i^{eq} V_i^{*g}. \tag{A.36}$$

The solvation volume is thus

$$\Delta V_s^* = V_s^{*l} - V_s^{*g} = \sum_i x_i^{eq} V_i^{*l} - \sum_i y_i^{eq} V_i^{*g}$$
$$= \sum_i x_i^{eq} \Delta V_i^* + \left[\bar{V}^g (x^{eq}) - \bar{V}^g (y^{eq}) \right]. \tag{A.37}$$

The meaning of Eq. (A.37) is as follows: the solvation volume ΔV_s^* is an average of the solvation volumes of the conformers (ΔV_i^*) with the distribution x^{eq}. In addition, the term in the square brackets is the change in the average pseudopartial moral volumes in the *gaseous* phase when we change from the distribution y^{eq} to the new distribution x^{eq}.

Another equivalent way of writing Eq. (A.37) is

$$\Delta V_s^* = \sum_i y_i^{eq} \Delta V_i^* + \left[\bar{V}^l (x^{eq}) - \bar{V}^l (y^{eq}) \right]. \tag{A.38}$$

The meaning of the two terms in Eq. (A.38) is similar to the meaning of the two terms in Eq. (A.37). In (A.37), the process of determining ΔV_s^* is to first change the distribution in the gaseous phase from y^{eq} to x^{eq} (see Figure A.3a), and then to solvate the solutes at the distribution x^{eq}. In (A.38), we first solvate the solutes at y^{eq}, and then change the distribution in the liquid from y^{eq} to x^{eq} [Figure A.3(b)].

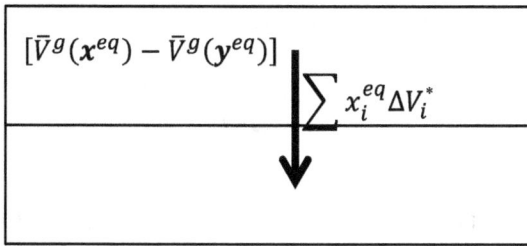

(a)

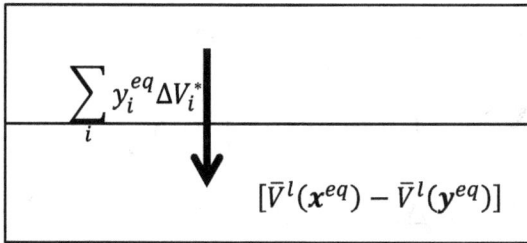

(b)

Fig. A.3. Two possible processes of solvation to determine the solvation volume.

A.5 The Solvation Enthalpy

Again, we start with Eq. (A.20) and take the temperature derivative to obtain

$$\exp\left(-\beta\mu_s^{*l}\right)\left(\mu_s^{*l} + TS_s^{*l}\right)$$

$$= \sum_i \exp\left(-\beta\mu_i^{*l}\right)\left(\mu_i^{*l} + TS_i^{*l}\right). \quad (A.39)$$

Using the equality $\mu_s^{*l} + TS_s^{*l} = H_s^{*l}$ and Eq. (A.18), we get

$$H_s^{*l} = \sum_i x_i^{\text{eq}} H_i^{*l}, \quad (A.40)$$

and similarly for the gaseous state,

$$H_s^{*g} = \sum_i y_i^{eq} H_i^{*g}. \qquad (A.41)$$

From Eqs. (A.40) and (A.41), we obtain the solvation enthalpy of the solute s:

$$\begin{aligned} \Delta H_s^* &= \sum_i x_i^{eq} H_i^{*l} - \sum_i y_i^{eq} H_i^{*g} \\ &= \sum_i x_i^{eq} \left(H_i^{*l} - H_i^{*g} \right) + \sum_i H_i^{*g} \left(x_i^{eq} - y_i^{eq} \right) \\ &= \sum_i x_i^{eq} \Delta H_i^* + \left[\bar{H}^g \left(x^{eq} \right) - \bar{H}^g \left(y^{eq} \right) \right]. \quad (A.42) \end{aligned}$$

As in Eq. (A.37), the solvation enthalpy is equal to the average solvation enthalpies of all the conformations plus the change in the enthalpy of the solute in the *gaseous* phase for changing the distribution from y^{eq} to x^{eq}.

Similarly, we can write Eq. (A.42) in an equivalent way:

$$\Delta H_s^* = \sum_i y_i^{eq} \Delta H_i^* + \left[\bar{H}^l \left(x^{eq} \right) - \bar{H}^l \left(y^{eq} \right) \right]. \quad (A.43)$$

Here, the average of ΔH_i^* is with the distribution y^{eq}, and the square brackets denote the change in the enthalpy in the *liquid phase*, when the distribution changes from y^{eq} to x^{eq}.

A.6 The Solvation Entropy

We found the relationship between $\Delta \mu_s^*$ and all the individual $\Delta \mu_i^*$. We also found the relationship between ΔH_s^* and all the individual ΔH_i^*. Therefore, we can in principle calculate

ΔS_s^* from

$$T \Delta S_s^* = -\Delta \mu_s^* + \Delta H_s^* \tag{A.44}$$

or, from Eq. (A.29),

$$\exp\left[\beta \left(\Delta H_s^* - T \Delta S_s^*\right)\right] = \sum_i x_i^{eq}$$
$$\times \exp\left[\beta \left(\Delta H_i^* - T \Delta S_i^*\right)\right], \tag{A.45}$$

$$\exp\left(-\beta T \Delta S_s^*\right) = \sum_i x_i^{eq} \exp\left(-\beta T \Delta S_i^*\right)$$
$$\times \exp\left[\beta \left(\Delta H_i^* - \Delta H_s^*\right)\right]. \tag{A.46}$$

We see that ΔS_s^* is not a simple average of ΔS_i^*; nor is $\exp\left(-\beta T \Delta S_s^*\right)$ a simple average of the quantities $\exp\left(-\beta T \Delta S_i^*\right)$.

Note that equation of the type A.43 can be written for the entropy of solvation as well.

Appendix B

The Definition of the Structure of Water and Changes in the Structure Induced by a Solute

The concept of the *structure of water* (SOW) has been in the literature for over a hundred years. Most authors *describe* the SOW rather than define it. For instance, Eisenberg and Kauzmann (1969) *described* how the SOW might look like in a microscopic snapshot of liquid water. They did not provide a definition of the SOW. To the best of my knowledge, the best way of defining the SOW is by using a mixture model approach to liquid water based on the water–water pair potential. [For details, see Ben-Naim (1992, 2009).]

We start with a pair potential of the form

$$U(X_1, X_2) = U_{LJ}(R_{12}) + U_{DD}(X_1, X_2) + \varepsilon_{HB} G(X_1, X_2),$$

(B.1)

where ε_{HB} is the hydrogen bond energy ($\varepsilon_{HB} \approx -6\,\text{kcal mol}^{-1}$) and $G(X_1, X_2)$ is essentially a geometric function, defined as

$$G(X_1, X_2) = \begin{cases} 1 & \text{whenever the configuration } X_1, X_2 \\ & \text{is favorable for forming an HB,} \\ \\ 0 & \text{otherwise.} \end{cases}$$

(B.2)

X_i is the vector describing the location and orientation of the ith water molecule.

The three terms in Eq. (B.1) correspond to the short-range Lennard–Jones (LJ) interaction, the long-range (dipole–dipole) and the intermediate range interaction (HBs) between the water-like particles. In order to formulate an analytical form for the function (X_1, X_2), it is convenient to think of a water molecule as having four "arms," i.e. four selected directions along which they can form HBs. These arms are along the four unit vectors pointing to the four vertices of a tetrahedron (Figure B.1). One can define four vectors: two along donor arm h_{i1} or h_{i2}, and two along acceptor arm l_{i1} or l_{i2}. In terms of these vectors, the function $G(X_1, X_2)$ is assumed to have the following general form:

$$G(X_i, X_j) = G_{\sigma'}(R_{ij} - R_H)$$

$$\times \left\{ \sum_{\alpha,\beta=1}^{2} G_\sigma \left[(h_{ia} \cdot u_{ij}) - 1 \right] G_\sigma \left[(l_{j\beta} \cdot u_{ij}) + 1 \right] \right.$$

$$\left. + \sum_{\alpha,\beta=1}^{2} G_{\sigma'}[(l_{i\alpha} \cdot u_{ij}) - 1]G_\sigma[(h_{j\beta} \cdot u_{ij}) + 1] \right\},$$

(B.3)

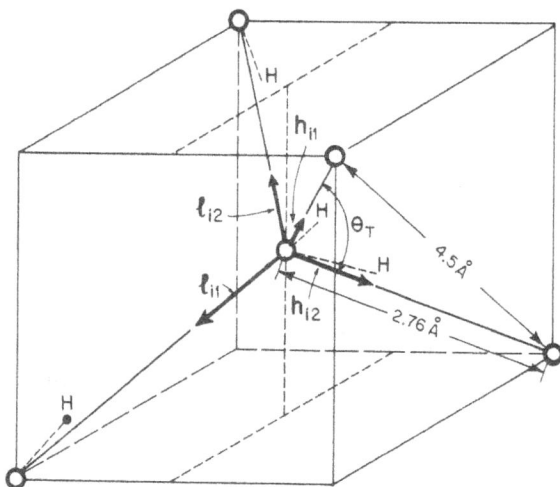

Fig. B.1. Geometry of the four "arms" of a water molecule.

where the function $G(x)$ may be chosen either as step function of the form

$$G_{\sigma'}(x) = \begin{cases} 1 & \text{for} \quad |x| < \sigma, \\ 0 & \text{for} \quad |x| \geq \sigma, \end{cases} \tag{B.4}$$

or as an unnormalized Gaussian function defined by

$$G_{\sigma'}(x) = \exp\left(\frac{-x^2}{2\sigma^2}\right). \tag{B.5}$$

The function (B.3), though cumbersome in appearance, is quite simple in content. Consider first the function $G_{\sigma'}(R_{ij} - R_H)$, where R_H is the intermolecular distance at which we expect an HB to be formed. A reasonable choice is $R_H = 2.76\,\text{Å}$. This function attains its maximal value of unity. It drops to zero — [either abruptly [Eq. (B.4)] or continuously [Eq. (B.5)] for $|R_{ij} - R_H| > \sigma'$. Next, we stipulate the relative orientation of the pair of molecules. The

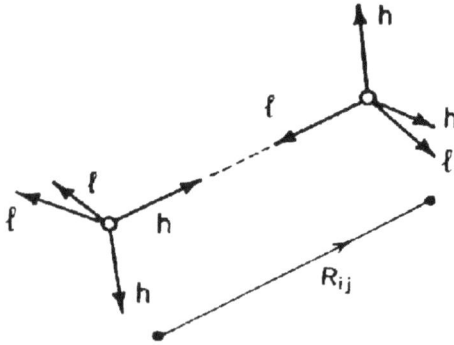

Fig. B.2. A favorable orientation of two water molecules to form a HB.

factor $G_\sigma[(\boldsymbol{h}_{i\alpha} \cdot \boldsymbol{u}_{ij}) - 1]$ attains its maximum value whenever the unit vector $\boldsymbol{h}_{i\alpha}$ (i.e. the donor arm $\boldsymbol{h}_{i\alpha}$, $\alpha = 1, 2$, of the ith molecule) is in the direction of the unit vector $\boldsymbol{u}_{ij} = \boldsymbol{R}_{ij}/R_{ij}$. Similarly, $G_\sigma[(\boldsymbol{l}_{j\beta} \cdot \boldsymbol{u}_{ij}) + 1]$ attains its maximum value whenever the direction of the acceptor arm $\boldsymbol{l}_{j\beta}$ is in the direction $-\boldsymbol{u}_{ij}$. Thus, the product of these three functions attains a value close to unity only if simultaneously R_{ij} is about R_H, the direction of $\boldsymbol{h}_{i\alpha}$ is about that of $\boldsymbol{u}_{ij}$, and the direction of $\boldsymbol{l}_{j\beta}$ is about that of $-\boldsymbol{u}_{ij}$. Such a configuration is said to be "favorable" for HB formation. Figure B.2 shows an example of such a configuration where one h arm forms an HB with an l arm. Clearly, if all of the above three conditions are fulfilled, then the interaction energy is about ε_{HB}. The sum of the various terms in the curly brackets of Eq. (A.3) arises from the total of eight possible favorable directions for HB formation (four when molecule i is a donor and four when molecule i is an acceptor). The variances σ and σ' are considered as adjustable parameters. Note that of the eight terms in the curly brackets, only one may be appreciably different from zero at any given configuration $X_i X_j$.

A choice of the square well function (A.4) leads to an on–off definition of an HB, i.e. two water molecules are either hydrogen-bonded or not. This simplified view is sometimes convenient in applications. The continuous definition of the HB potential using Eq. (B.5) provides a gradual change from a hydrogen-bonded to a non-hydrogen-bonded pair of water molecules. This is clearly a more realistic view that can be described as stretching and bending of the HB.

The concept of the SOW is ubiquitous in the literature on water and aqueous solutions. In most cases, this concept is discussed in a very qualitative way. For instance, Eisenberg and Kauzmann (1969) start from the concept of "structure" as applied to the crystal ice I_h, then proceed to describe the structure of liquid water. Most of the discussion is descriptive, and no quantitative definition of the SOW is offered.

In the theory of liquids, the *structure* of the liquid is usually defined in terms of the molecular distribution function. This definition, though quantitative, is not satisfactory when applied to liquid water. What is needed is a *number* that measures the extent of the SOW, as it is currently understood, with the help of which one can compare the structures of the liquid in two different states. We present here one possible definition which is based on the form of the pair potential which conforms to what is expected from a measure of the SOW, and one that can be computed by any simulation technique.

Using the definition of the pair potential in Eq. (B.1), we define the following function:

$$\psi_i(X^N) = \sum_{\substack{j=1 \\ j \neq i}}^{N} G(X_i, X_j). \tag{B.6}$$

For each configuration X^N of the N water molecule, we choose a particular molecule, — say, the ith one. Since $G(X_i, X_j)$ is unity whenever molecule j is in a favorable configuration to form an HB with i, the sum on the RHS of Eq. (B.6) *counts* all the water molecules that are hydrogen-bonded to i at a specific configuration of the entire system. As we have chosen σ in Eq. (B.3) to be small enough so that two water molecules can form at most one HB between them, the value of $\psi_i(X^N)$ can roughly change between zero and four.

We now define the average value of $\psi(X^N)$ in, say the T, V, N ensemble as

$$\langle \psi = \int dX^N P(X^N) \psi_1 \left(X^N \right). \qquad (B.7)$$

This is the average number of HBs formed by a specific water molecule (say, a molecule numbered 1), in a system of pure water at T, V, N. Similar definitions apply to the T, P, N ensemble.

The quantity $\langle \psi \rangle$ is related to the total average number of HBs in the system, which we denote $\langle N_{HB} \rangle$. This is defined as

$$\langle N_{HB} \rangle = \int dX^N P \left(X^N \right) \sum_{i=1}^{N} \sum_{\substack{j=1 \\ i \neq j}}^{N} G \left(X_i, X_j \right)$$

$$= \frac{1}{2} \int dX_1 dX_2 \rho^{(2)}(X_1, X_2) G(X_1, X_2). \qquad (B.8)$$

Here, $\rho^{(2)}(X_1, X_2)$ is the pair distribution function.

Since the sum over all i and j (with $i \neq j$) counts the number of all different pairs of molecules, the integral is the average number of pairs of molecules, the configuration of which is within the favorable range for forming an HB. The factor $\frac{1}{2}$ is included because the sum in the integrand counts each pair twice.

The second form on the RHS of Eq. (B.8) converts the average into an integral over the pair distribution function. The relation between $\langle N_{HB} \rangle$ and $\langle \psi \rangle$ is obtained from Eqs. (B.6)–(B.8), i.e.

$$\langle N_{HB} \rangle = \frac{1}{2} \int dX^N P\left(X^N\right) \sum_{i=1}^{N} \sum_{\substack{j=1 \\ i \neq j}}^{N} G\left(X_i, X_j\right)$$

$$= \frac{1}{2}N \int dX^N P\left(X^N\right) \psi_1\left(X^N\right)$$

$$= \frac{1}{2}N\langle \psi \rangle. \tag{B.9}$$

Thus, either $\langle \psi \rangle$ or $\langle N_{IIB} \rangle$ can serve as a definition of the SOW. Clearly, this definition conforms with our intuitive expectation of a measure of the SOW. If $\langle \psi \rangle \approx 4$, each molecule is on the average hydrogen-bonded to four other molecules, as in ice I_h. This is a highly structured state. On the other hand, $\langle \psi \rangle \approx 0$ corresponds to the lowest structured state.

We can use the definition of $\psi(X^N)$ in Eq. (B.5) to construct an exact *mixture model* approach to liquid water:

$$N_n\left(X^N\right) = \sum_{i=1}^{N} \delta\left[\psi_i\left(X^N\right) - n\right], \tag{B.10}$$

where δ can be either a Kronecker delta or a Dirac delta function, depending on whether we choose the discrete or continuous definition function of the HB [i.e. either Eq. (B.4) or Eq. (B.5)]. $N_n\left(X^N\right)$ is the number of molecules which form n HBs when the system is at a specific configuration X^N. The average number of such molecules is

$$\langle N_n \rangle = \int dX^N P\left(X^N\right) \sum_{i=1}^{N} \delta\left[\psi_i\left(X^N\right) - n\right]$$

$$= N \int dX^N P\left(X^N\right) \delta\left[\psi_i\left(X^N\right) - n\right].$$

(B.11)

From Eq. (B.11), we can define the mole fraction x_n corresponding to $\langle N_n \rangle$, i.e.

$$x_n = \frac{\langle N_n \rangle}{N}.$$

(B.12)

Note that once we have defined the mole fractions of the various species, we have in fact constructed a mixture model approach to liquid water. A more general procedure for constructing a mixture model is described in Ben-Naim (1992, 2009).

The average number of HBs in the system is thus

$$\frac{N}{2} \sum_{n=0}^{4} n x_n = \frac{N}{2} \int dX^N P\left(X^N\right) \sum_{n=0}^{4} \delta\left[\psi_i\left(X^N\right) - n\right]$$

$$= \frac{N}{2} \int dX^N P\left(X^N\right) \psi_i\left(X^N\right) = \langle N_{\mathrm{HB}} \rangle,$$

(B.13)

which is the same as $\langle N_{\mathrm{HB}} \rangle$ in Eq. (B.8).

Having defined the concept of the SOW, we next ask how we can measure this structure. The following is an approximate relation between the solvation Gibbs energy of water and the quantity $\langle \psi \rangle$ or, equivalently, $\langle N_{HB} \rangle$ in pure water. This approximation is based on the assumption that H_2O and D_2O may each be represented by a molecular model with a pair potential of the form (B.1). The two liquids are assumed to have the same pair potential except for a difference in the HB energy ε_{HB}, which we denote ε_D and ε_H for D_2O and H_2O respectively. In other words, we assume that the geometry and the LJ parts of the interaction are the same for H_2O and D_2O, and only the strength of the HB energies is different.

Starting with the expression for the solvation Gibbs energy of water in pure water,

$$\Delta G_W^* = -k_B T \ln \langle \exp(-\beta B_W) \rangle_0, \qquad \text{(B.14)}$$

where B_W is the binding energy of an added water molecule at a configuration X_0 to a system of N water molecules at a fixed configuration $X^N = X_1, \ldots, X_N$. The average in Eq. (B.14) is over all configurations of the N water molecules in the T, P, N ensemble.

We differentiate ΔG_W^* with respect to ε_{HB} at constant T, P, N to obtain

$$\left(\frac{\partial \Delta G_W^*}{\partial \varepsilon_{HB}} \right)_{T,P,N} = -k_B T \frac{\partial}{\partial \varepsilon_{HB}}$$

$$\times \left\{ \ln \frac{\int dV \exp(-\beta PV) \int dX^N \exp(-\beta U_N - \beta B_W)}{\int dV \exp(-\beta PV) \int dX^N \exp(-\beta U_N)} \right\}$$

$$= \frac{1}{2} \int dV \exp\left(-\beta PV\right) \int dX^N P\left(X^N | X_0\right)$$

$$\times \sum_{i=0}^{N} \sum_{\substack{j=0 \\ j \neq 1}}^{N} G(i,j)$$

$$= \langle N_{\mathrm{HB}} \rangle_W^{(N+1)} - \langle N_{\mathrm{HB}} \rangle_0^{(N)}. \tag{B.15}$$

Note that in the last form on the RHS of Eq. (B.15) we have two different averages in the T, P, N ensemble. The first is the *conditional* average number of HBs in a system of $N + 1$ water molecules, given that one water molecule is at a fixed configuration X_0. The second is the average number of HBs of a system of N water molecules. The latter is the same as in Eq. (B.8) except that the average is taken in the T, P, N ensemble. The subscripts 0 and w in these averages indicate that the averages are taken with respect to the distributions $P\left(V, X^N\right)$ and $P\left(V, X^N | X_0\right)$, respectively.

If we release the constraint on the added water molecule to be at the specific configuration X_0, the average number of HBs should not change. Therefore, the difference in Eq. (B.15) may be written as

$$\left(\frac{\partial \Delta G_W^*}{\partial \varepsilon_{\mathrm{HB}}}\right)_{T,P,N} = \langle N_{\mathrm{HB}} \rangle_0^{(N+1)} - \langle N_{\mathrm{HB}} \rangle_0^{(N)}$$

$$= \frac{N+1}{2} \langle \Psi \rangle_0 - \frac{N}{2} \langle \Psi \rangle_0$$

$$= \frac{1}{2} \langle \Psi \rangle_0. \tag{B.16}$$

Thus, the derivative of ΔG_W^* with respect to the parameter ε_{HB} is a measure of the SOW.

We can now use Eq. (B.16) to estimate the SOW. The assumption is made that $\varepsilon_D - \varepsilon_H$ is small enough so that we can expand ΔG_W^* to first order in $\varepsilon_D - \varepsilon_H$ to obtain

$$\Delta G_{D_2O}^* - \Delta G_{H_2O}^* \approx \frac{1}{2} \langle \psi \rangle_0 (\varepsilon_D - \varepsilon_H). \quad (B.17)$$

This relation may be used to estimate the extent of structure $\langle \psi \rangle_0$ from experimental data on the solvation Gibbs energies, and from $\varepsilon_D - \varepsilon_H$. [For details, see Ben-Naim (2009).]

Similarly, by differentiating the solvation Gibbs energy of the solute s, ΔG_S^*, with respect to the parameter ε_{HB}, we obtain

$$\left(\frac{\partial \Delta G_S^*}{\partial \varepsilon_{HB}} \right)_{T,P,N} = \langle N_{HB} \rangle_S - \langle N_{HB} \rangle_0 , \quad (B.18)$$

where, on the RHS, we have the change in the average number of HBs caused by placing s at a fixed position in the solvent.

Assuming that $\varepsilon_D - \varepsilon_H$ is small enough, we can expand ΔG_S^* to first order in $\varepsilon_D - \varepsilon_H$ to obtain

$$\Delta G_S^* (D_2O) - \Delta G_S^* (H_2O) = \left(\langle N_{HB} \rangle_S - \langle N_{HB} \rangle_0 \right)$$
$$\times (\varepsilon_D - \varepsilon_H). \quad (B.19)$$

Thus, from the knowledge of the solvation Gibbs energies of a solute s in H_2O and in D_2O, we can estimate the quantitative change in the SOW induced by the solutes. [More details can be found in Ben-Naim (2009).]

Appendix C

Le Chatelier's Principle Applied to the Two-Structure Model for Water

Structural changes in the solvent is a response process triggered by change in the environment. One changes a parameter in the environment, such as temperature T, pressure P or number of solute molecules N_s, and observe a resulting change in the relative change in the concentrations of the *low-local-density* (LLD) and the *high-local-density* (HLD) forms of the water molecule (w).

The quantitative change in the concentration of the LLD, x_L, caused by changes in T, P or N_s, is obtained as follows.

Consider a system at temperature T and pressure P, having N_w water molecules w, and N_s solute molecules s. If N_L and N_H are the average number of L and H molecules, respectively, then $N_w = N_L + N_H$.

For the purpose of this section, it does not matter how we have defined the two forms of water molecules. Let N_L and N_H be the average number of molecules of types L and H, at equilibrium:

$$L \rightleftarrows H. \qquad (C.1)$$

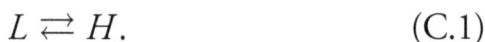

The condition of equilibrium is

$$\mu_L = \mu_H. \qquad (C.2)$$

Viewing both μ_L and μ_H as a function of the variables (T, P, N_w, N_s), we have the implicit equation

$$\mu_L(T, P, N_w, N_s; N_L) = \mu_H(T, P, N_w, N_s; N_L) \qquad (C.3)$$

which can be solved for N_L for any given constants (T, P, N_w, N_s).

We define the mole fraction of L as

$$x_L = \frac{N_L}{N_w}. \qquad (C.4)$$

Hence, also, $x_H = 1 - x_L$. Note that the mole fractions are defined with respect to the total number of water molecules, and not with respect to the total number of molecules in the system. Note also that in Eq. (C.3) N_L is not an independent variable. It is dependent on the parameters T, P, N_w, N_s at equilibrium. On the other hand, we might "freeze in" the reaction (C.1) so that the conversion between the two forms L and H is precluded. In this case, the independent variables are T, P, N_s, N_L, N_H (with $N_w = N_L + N_H$).

Define the difference:

$$\Delta\mu = \mu_L - \mu_H. \qquad (C.5)$$

For any arbitrary change in the variables T, P, N_s, N_L, N_H, we can write the total differential of $\Delta\mu$ as

$$d\Delta\mu = \left(\frac{\partial\Delta\mu}{\partial T}\right)_* dT + \left(\frac{\partial\Delta\mu}{\partial P}\right)_* dP + \left(\frac{\partial\Delta\mu}{\partial N_s}\right)_* dN_s$$

$$+ \left(\frac{\partial\Delta\mu}{\partial N_L}\right)_* dN_L + \left(\frac{\partial\Delta\mu}{\partial N_H}\right)_* dN_H. \qquad \text{(C.6)}$$

Note that all the derivatives on the RHS of Eq. C.6 are taken in the frozen-in system, viewing $\Delta\mu$ as a function of the independent variables T, P, N_s, N_L, N_H. These derivatives are denoted by the asterisk as a subscript.

Next, we take the derivatives of $\Delta\mu$ along the equilibrium line. This means that we follow the change in $\Delta\mu$ while maintaining the equilibrium condition (C.2). These derivatives will be denoted by the subscript "eq".

The three derivatives of importance are with respect to T, P and N_s. Taking the derivative of $\Delta\mu$ with respect to the temperature, we get from Eq. (C.6)

$$0 = \left(\frac{\partial\Delta\mu}{\partial T}\right)_{eq}$$

$$= \left(\frac{\partial\Delta\mu}{\partial T}\right)_* + \left[\left(\frac{\partial\Delta\mu}{\partial N_L}\right)_* - \left(\frac{\partial\Delta\mu}{\partial N_H}\right)_*\right]\left(\frac{\partial N_L}{\partial T}\right)_{eq}. \qquad \text{(C.7)}$$

Note that since the derivative on the LHS of Eq. (C.7) is taken along the *equilibrium line* (eq), it is zero. Also, maintaining the equilibrium condition implies that $dN_L + dN_H = 0$.

The quantity in the squared brackets on the RHS of Eq. (C.7) is denoted

$$\mu_{**} \equiv \left(\frac{\partial \Delta \mu}{\partial L}\right)_* - \left(\frac{\partial \Delta \mu}{\partial N_H}\right)_*$$
$$= \mu_{LL} - 2\mu_{LH} + \mu_{HH} > 0, \qquad (C.8)$$

where $\mu_{ij} = \frac{\partial^2 G}{\partial N_i \partial N_j} = \mu_{ji}$ is the second derivative of the Gibbs energy. The quantity μ_{**} defined in (C.8) is always positive. This follows from the condition of stability of the system; see Appendix D.

At equilibrium, we have

$$\Delta \mu = \Delta H - T \Delta S, \qquad (C.9)$$

where $\Delta H = \bar{H}_L - \bar{H}_H$ and $\Delta S = \bar{S}_L - \bar{S}_H$. We also have the identity

$$\left(\frac{\partial \Delta \mu}{\partial T}\right)_* = -\Delta S = -\frac{\Delta H}{T}. \qquad (C.10)$$

Therefore, the equality (C.7) may be rewritten as

$$\left(\frac{\partial N_L}{\partial T}\right)_{eq} = \mu_{**}^{-1} \Delta S = \mu_{**}^{-1} \frac{\Delta H}{T}. \qquad (C.11)$$

This relationship is a particular example of Le Chatelier's principle. Increasing the temperature ($dT > 0$) will cause an increase (or decrease) in N_L if and only if ΔH is positive (or negative).

Taking a derivative of $\Delta\mu$ with respect to pressure, along the equilibrium line, leads to the identity

$$0 = \left(\frac{\partial\Delta\mu}{\partial P}\right)_* + \mu_{**}\left(\frac{\partial N_L}{\partial P}\right)_{eq} \qquad (C.12)$$

or, equivalently,

$$\left(\frac{\partial N_L}{\partial P}\right)_{eq} = -\mu_{**}^{-1}\left(\frac{\partial\Delta\mu}{\partial P}\right)_* = -\mu_{**}^{-1}\Delta V, \qquad (C.13)$$

where $\Delta V = \bar{V}_L - \bar{V}_H$ is the volume change in the process (C.1).

Similarly, taking the derivative with respect to N_s along the equilibrium line leads to the identity

$$\left(\frac{\partial N_L}{\partial N_s}\right)_{eq} = -\mu_{**}^{-1}\left(\frac{\partial\Delta\mu}{\partial N_s}\right)_*. \qquad (C.14)$$

Note that the derivative on the RHS of Eq. (C.14) is taken in a frozen-in state, whereas the derivative on the LHS is taken along the equilibrium line (eq).

Thus, we have the following three fundamental equations:

$$\left(\frac{\partial x_L}{\partial T}\right)_{eq} = -\mu_{**}^{-1}\left(\frac{\partial\Delta\mu}{\partial T}\right)_*, \qquad (C.15)$$

$$\left(\frac{\partial x_L}{\partial P}\right)_{eq} = -\mu_{**}^{-1}\left(\frac{\partial\Delta\mu}{\partial P}\right)_*, \qquad (C.16)$$

$$\left(\frac{\partial x_L}{\partial N_s}\right)_{eq} = -\mu_{**}^{-1}\left(\frac{\partial\Delta\mu}{\partial N_s}\right)_*. \qquad (C.17)$$

These equations were used within the exact mixture model approach to liquid water [Ben-Naim (2009)]. There are no requirements on any ideality condition of the mixture. The same equations were used in the theory of protein folding. See [Ben-Naim (2013)].

Appendix D

The Stability Condition
for a Chemical Reaction: $A \rightleftarrows B$

We show that the quantity $\mu_{AA} - 2\mu_{AB} + \mu_{BB}$ is always positive. This follows from the condition of stability of the system. At equilibrium, at constant T, P, N_S, N_W, the Gibbs energy has a minimum with respect to variation in N_A and N_B. Thus, for any variation in N_A away from equilibrium, the corresponding change in the Gibbs energy must be positive. Hence, expanding G to second order in dN_A, we get

$$
\begin{aligned}
0 \le dG &= G(N_A + dN_A, N_B + dN_B) - G(N_A, N_B) \\
&= \frac{\partial G}{\partial N_A} dN_A + \frac{\partial G}{\partial N_B} dN_B \\
&\quad + \frac{1}{2}\left[\frac{\partial^2 G}{\partial N_A^2} dN_A^2 + 2\frac{\partial^2 G}{\partial N_A \partial N_B} dN_A dN_B \right. \\
&\quad \left. + \frac{\partial^2 G}{\partial N_B^2} dN_B^2 \right] + \cdots .
\end{aligned}
\tag{D.1}
$$

At equilibrium, the condition of chemical equilibrium leads to

$$\frac{\partial G}{\partial N_A} dN_A - \frac{\partial G}{\partial N_B} dN_A = (\mu_A - \mu_B) dN_A = 0. \quad \text{(D.2)}$$

Therefore, in the limit of small dN_A, we must have

$$\frac{1}{2} [\mu_{AA} - 2\mu_{AB} + \mu_{BB}] dN_A^2 \geq 0 \quad \text{(D.3)}$$

or, equivalently,

$$\mu_{AA} - 2\mu_{AB} + \mu_{BB} \geq 0. \quad \text{(D.4)}$$

Appendix E

Solvation Gibbs Energy of the Hard Sphere and the Work of Cavity Formation

In this appendix, we briefly show that the expression for $\Delta\mu_S^*$ for any solute in any solvent is

$$\Delta\mu_S^* = -k_B T \ln \langle \exp(-\beta B_S) \rangle_0, \qquad \text{(E.1)}$$

and that this expression follows directly from the definition of the solvation process. Then we apply the expression to a hard sphere (HS) solute.

Let $\Delta(T, P, N)$ be the T, P, N partition function (PF) of a system with composition $N = \{N_1, \ldots, N_c\}$ and the total number of molecules $N = \sum N_1$. Let $\Delta(T, P, N; R_s)$ be the PF of the same system, but with an additional molecule s placed at a fixed position R_s. The molecule s could be the same as one of the other N molecules, or a different one. The pseudochemical potential is defined by

$$\exp\left(-\beta\mu_s^{*l}\right) = \frac{\Delta(T,P,N;R_s)}{\Delta(T,P,N)}$$

$$= q_s \int dV dX^N P\left(V,X^N\right)\exp(-\beta B_s)$$

$$= q_s \left\langle \exp\left(-\beta B_S\right)\right\rangle_0, \qquad (E.2)$$

where $P(V,X^N)$ is the probability density of finding the volume V and the configuration X^N in the T,P,N ensemble, and B_S is the *binding energy* of s, defined by

$$B_s = U\left(R_s,X_1,\ldots,X_N\right) - U\left(X_1,\ldots,X_N\right). \qquad (E.3)$$

Note that the average in Eq. (E.2) is taken with the distribution $P(V,X^N)$ of the "pure" solvent, i.e. the system of N molecules *before* the insertion of s at R_s. q_S is the internal PF of s, which is assumed to be separable from the configurational PF.

Assuming that q_S is the same in the liquid and in an ideal gas phase, the pseudochemical potential of s in an ideal gas phase is simply

$$\exp\left(-\beta\mu_S^{*ig}\right) = q_S. \qquad (E.4)$$

Hence, from Eqs. (E.2) and (E.4), we get

$$\Delta\mu_S^* = -k_B T \ln\left\langle \exp\left(-\beta B_S\right)\right\rangle_0. \qquad (E.5)$$

A special case of (E.5) is for a HS solute. In this case, the interaction between s and any other molecule is either zero or infinity, according to whether the molecule is away from or within the excluded region produced by s. The function

$\exp\left(-\beta B_{HS}\right)$ is accordingly either zero or one. Therefore, the average in Eq. (E.5) may be written as

$$\int dV\,dX^N P\left(V, X^N\right) \exp\left(-\beta B_s\right)$$

$$= \int_{V - V^{EX}(R_s)} dV\,dX^N P\left(V, X^N\right)$$

$$= P\left(V^{EX}, R_s\right), \tag{E.6}$$

where $V - V^{EX}(R_s)$ is the region for which $\exp\left(-\beta B_S\right)$ is unity. Since $P(V, X^N)$ is a probability density, the integral on the RHS of Eq. (E.6) is the probability of finding all particles in the region $V - V^{EX}(R_s)$ or, equivalently, the probability that no solvent particle will occupy the region $V^{EX}(R_s)$. This is the same as the probability of finding an empty cavity of volume V^{EX} at R_s which we denote $P(V^{EX}, R_s)$. Thus, for the HS solute we have

$$\Delta\mu^*_{HS} = -k_B T \ln P(V^{EX}, R_s). \tag{E.7}$$

It should be noted that this relation is valid for any choice of the location R_s. However, R_s is included in the notation on the RHS of Eq. (E.7) to stress that cavity must be at some *fixed* position in the liquid. The probability of finding a cavity at a specific location R_s is different from the probability of finding the cavity of the same size *anywhere* in the system.

A special case of (E.7) is for a point HS. A point HS is defined as a HS of radius 0. For a one-component solvent with an effective diameter σ, the size of the excluded volume formed by the point HS is simply $\frac{4\pi\sigma^3}{3}$ (see Figure E.1).

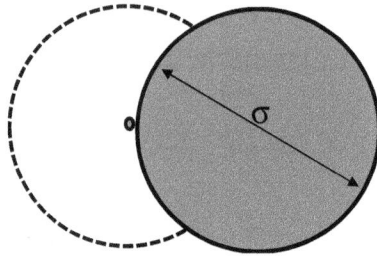

Fig. E.1. The excluded volume above a hard-point particle.

The sphere of volume $\frac{4\pi\sigma^3}{3}$ at the fixed position R_s can accommodate at most one solvent particle at any given time. Therefore, the probability of finding the volume $\frac{4\pi\sigma^3}{3}$ at R_s empty is

$$P_0(V^{\text{EX}} = \frac{4\pi\sigma^3}{3}, R_s) = 1 - \rho\frac{4\pi\sigma^3}{3}, \qquad \text{(E.8)}$$

and the corresponding solvation Gibbs energy is

$$\Delta\mu^*_\bullet = -k_B T \ln\left(1 - \rho\frac{4\pi\sigma^3}{3}\right). \qquad \text{(E.9)}$$

Thus, the solvation Gibbs energy of a hard point is determined by the diameter of the solvent molecules and by the density of the solvent. Here, we assume for simplicity that the solvent contains only one kind of molecules. [For more details, see Ben-Naim (2006).]

Appendix F

Formal Expression
for the Thermodynamic Quantities
of Solvation of a Nonpolar Solute

We start with the expression for the solvation Helmholtz energy of the solute s in a solvent w.

$$\Delta A_s^* = -k_B T \ln \left\langle \exp\left(-\beta B_s\right)\right\rangle_0 . \qquad \text{(F.1)}$$

In Eq. (F.1), the average with the probability distribution for the configurations of the solvent molecules *before* inserting the solute in a fixed position in the solvent.

$$P_0(X^N) = \frac{\exp\left[-\beta U_N\left(X^N\right)\right]}{\int \cdots \int dX^N \exp\left[-\beta U_N\left(X^N\right)\right]}. \qquad \text{(F.2)}$$

B_s is the total binding energy of s to all the solvent molecules. Here, we assume that the solvent consists of N water molecules and we insert one solute s at a fixed position R_s:

$$B_s = \sum_{i=1}^{N} U(R_s, X_i). \qquad \text{(F.3)}$$

For the interpretation of the entropy and the enthalpy of solvation in terms of structural changes in the solvent, it will be convenient to rewrite the average in Eq. (F.1) using the distribution of the binding energy.

We first define the distribution function:

$$P(v)dv = dvP(X^N)\delta[B_s - v]dX^N, \qquad (F.4)$$

where $\delta[B_s - v]$ is the Dirac delta function. The quantity $P(v)dv$ is the probability of finding the binding energy B_s, between v and $v + dv$. This is also the mole fraction of solute molecules (in a dilute solution in w) having a binding energy between v and $v + dv$.

With the distribution function defined in Eq. (F.4), we can rewrite the solvation Helmholtz energy of the solute as

$$\langle \exp(-\beta B_s) \rangle_0 = \int \cdots \int dX^N P_0\left(X^N\right) \exp\left(-\beta B_s\right)$$

$$= \sum_{v_i} \int_{\Delta v_i \leq B_s \leq v_i + \Delta v} \cdots$$

$$\times \int dX^N P_0\left(X^N\right) \exp\left(-\beta B_s\right)$$

$$\approx \sum_{v_i} dv \int \cdots \int dX^N P_0\left(X^N\right)$$

$$\times \exp\left(-\beta B_s\right) \delta\left(B_s - v_i\right). \qquad (F.5)$$

In the first step on the RHS of Eq. (F.5), we rewrite the integral over all configurations of the solvent molecules as a sum over subspaces, each of which is characterized by the condition $v_i \leq B_s \leq v_i + \Delta v$. In the limit of infinitesimal intervals Δv, we can rewrite each of the integrals over the subspace as an integral over the entire configurational space with the added delta function to the integrand. In the limit $dv \to 0$, we can convert the sum on the RHS of Eq. (F.5) into an integral, i.e.

$$\langle \exp(-\beta B_s) \rangle_0 = \int_{-\infty}^{\infty} dv P(v) \exp(-\beta v). \qquad (F.6)$$

Next, we take the temperature derivative of ΔA_s^* in Eq. (F.1) to obtain

$$
-\Delta A_s^* = \left(\frac{\partial \Delta A_s^*}{\partial T} \right)_{T,V}
$$
$$
= k_B \ln \langle \exp(-\beta B_s) \rangle_0
$$
$$
= \frac{1}{T} \left[\frac{\int \cdots \int dX^N U_N \exp(-\beta U_N))}{\int \cdots \int dX^N \exp(-\beta U_N)} \right.
$$
$$
\left. - \frac{\int \cdots \int dX^N [U_N + B_s] \exp[-\beta(U_N + B_s)]}{\int \cdots \int dX^N \exp[-\beta(U_N + B_s)]} \right]
$$
$$
= \frac{1}{T} \left[\langle U_N \rangle_0 - \langle U_N \rangle_s + \langle B_s \rangle_s \right]. \qquad (F.7)
$$

Hence, we can write the solvation entropy and the solvation energy as

$$
\Delta A_s^* = k_B \ln \exp(-\beta B_s)))_0
$$
$$
+ \frac{1}{T} \left[\langle B_s \rangle_s + (\langle U_N \rangle_s - \langle U_N \rangle_0) \right] \qquad (F.8)
$$
$$
\Delta E_s^* = \Delta A_s^* + T \Delta A_s^*
$$
$$
= \langle B_s \rangle_s + (\langle U_N \rangle_s - \langle U_N \rangle_0). \qquad (F.9)
$$

Note that the symbol $\langle\,\rangle_0$ stands for an average over all configurations of the solvent molecules with the distribution of $P_0(X^N)$ defined in Eq. (F.2). On the other hand, the symbol $\langle\,\rangle_s$ refers to an average with the conditional distribution

$$P\left(X^N\,\middle|\,R_s\right) = \frac{\exp\left(-\beta U_N - \beta B_s\right)}{\int dX^N \exp\left(-\beta U_N - \beta B_s\right)}. \qquad \text{(F.10)}$$

Note that both ΔA_s^* and ΔE_s^* contain the term $\langle U_N\rangle_s - \langle U_N\rangle_0$. This term may be interpreted in terms of structural changes induced by the solute s on the solvent. (See Appendix G).

Appendix G

The Solvation Entropy and Solvation Energy; Structural Changes in the Solvent

We have seen in Appendix F that the solvation entropy and solvation energy of a solute s can be written as

$$\Delta S_s^* = k_B \ln \langle \exp(-\beta B_s) \rangle_0$$
$$+ \left[\langle B_s \rangle_s + (\langle U_N \rangle_s - \langle U_N \rangle_0) \right],$$
$$\Delta E_s^* = \langle B_s \rangle_s + (\langle U_N \rangle_s - \langle U_N \rangle_0). \qquad (G.1)$$

We focus here on the term $\langle U_N \rangle_s - \langle U_N \rangle_0$ which is common to ΔS_s^* and ΔE_s^*. (Here, we work in the T, V, N ensemble, but the same is true in the T, P, N ensemble [see Ben-Naim (2006, 2009)]. This quantity is simply the *change* in the total interaction energy among the solvent molecules caused by the insertion of a solute s at some fixed position in the solvent.

We now want to reinterpret this quantity in terms of structural changes in the solvent. There are many ways of doing this, depending on how one chooses to define the "structure" of the solvent.

As we have done in Appendix F, we rewrite both $\langle U_N \rangle_0$ and $\langle U_N \rangle_s$ in terms of the binding energy distribution function for the solvent molecules. This distribution function defines quasicomponents in the liquid. Therefore, $\langle U_N \rangle_s - \langle U_N \rangle_0$ may be interpreted as the change in the distribution of quasicomponents characterized by the binding energies of the solvent molecules. The more general case is discussed in Ben-Naim (2009). Here, we discuss a simple case of a two-component model.

Suppose that each solvent molecule can be in one of two "states"; it can have either binding energy v_1 or binding energy v_2. The distribution of binding energies in this case is simply

$$P(v) = x_1 \delta(v - v_1) + x_2 \delta(v - v_2), \qquad \text{(G.2)}$$

where δ is the Dirac delta function, and x_1 and x_2 are the mole fractions of the two components. With this distribution function we can rewrite the difference $\langle U_N \rangle_s - \langle U_N \rangle_0$ as

$$\langle U_N \rangle_s - \langle U_N \rangle_0 = \frac{N}{2}(v_1 - v_2)\left(\frac{\partial x_1}{\partial N_s}\right)_{T,V}. \qquad \text{(G.3)}$$

If the mole fraction x_1 (of the component having binding energy v_1) represents the "structure" of the solvent, then the whole term on the RHS of Eq. (G.3) is the contribution of the change in the "structure" of the solvent to the solvation energy of the solute. A more detailed application of the two structure model is represented in Ben-Naim (2009).

Appendix H

The Solubility of a Simple Solute in Water and Structural Changes Induced in the Solvent

In this appendix, we use a very simple two-structure model to describe the effect of structural changes induced by a solute or by a cosolvent on the solubility of the solute.

The chemical potential of a solute s is defined by

$$\mu_S = \left(\frac{\partial G}{\partial N_s} \right)_{T,P,N_w}. \tag{H.1}$$

We view the solvent water as a mixture of two components, L and H. Let N_L and N_H be the average number of L and H at the given temperature T and pressure P, for a system of composition N_w, N_S. The quantities N_L and N_H are determined by the condition of chemical equilibrium

$$\mu_L(T, P, N_L, N_H, N_S) = \mu_H(T, P, N_L, N_H, N_S), \tag{H.2}$$

with the condition $N_L + N_H = N_w$, N_w being the total number of water molecules. Viewing G as a function of

311

the independent variables T, P, N_L, N_H, we can write the chemical potential (T and P are always constant, and will be omitted from the notation):

$$\mu_S = \left(\frac{\partial G}{\partial N_S}\right)_{N_w} = \left(\frac{\partial G}{\partial N_S}\right)_{N_L, N_H} + \left(\frac{\partial G}{\partial N_L}\right)_{N_S N_H}$$

$$\times \left(\frac{\partial N_L}{\partial N_S}\right)_{N_w} + \left(\frac{\partial G}{\partial N_H}\right)_{N_S N_L} \left(\frac{\partial N_H}{\partial N_S}\right)_{N_w}$$

$$= \mu_S^{\text{fr}} + (\mu_L - \mu_H)\left(\frac{\partial N_L}{\partial N_S}\right)_{N_w} = \mu_S^{\text{fr}}. \qquad \text{(H.3)}$$

The last form of the equality (H.3) follows from the condition of chemical equilibrium (H.2).

The conclusion derived from (H.3) may be stated as follows.

The chemical potential of s in the system is not affected by the structural changes induced by s. Note that any other partial molar quantity might be affected by the structural changes induced by s.

It should also be noted that the above conclusion is valid for any initial state (T, P, N_w, N_S), when the "freeze-in" of the equilibrium $L \leftrightarrows H$ is imposed on this state.

Next, we explore a different situation. Suppose that we start with pure water at (T, P, N_w) or, equivalently, at (T, P, N_L^0, N_H^0), where N_L^0 and N_H^0 are determined by the condition of chemical equilibrium (H.2) but with $N_S = 0$. We now freeze-in the equilibrium $L \rightleftarrows H$, and add a *finite* quantity of s, N_S. The chemical potential in this frozen-in state is $\mu_S(T, P, N_L^0, N_H^0, N_S)$. Next, we release the constraint on fixed N_L^0, N_H^0 and ask: In which direction will the chemical potential of s change?

To first order in $\Delta N_L = N_L - N_L^0$, we write the expansion of the chemical potential:

$$\mu_S(T, P, N_L, N_H, N_S)$$

$$= \mu_S^{\text{fr}}(T, P, N_L^0, N_H^0, N_S) + \left(\frac{\partial \mu_S}{\partial N_L} - \frac{\partial \mu_S}{\partial N_H} \right) \Delta N_L$$

$$= \mu_S^{\text{fr}} + \frac{\partial(\mu_L - \mu_H)}{\partial N_S} \Delta N_L. \qquad (H.4)$$

Note that ΔN_L is a result of the addition of the *finite* quantity of the solute s. The derivative $\partial(\mu_L - \mu_H)/\partial N_S$ is related to the structural changes in the solvent by the identity (see Appendix C)

$$\left(\frac{\partial(\mu_L - \mu_H)}{\partial N_S} \right)_{N_L^0 N_H^0}$$

$$= -(\mu_{LL} - 2\mu_{LH} + \mu_{HH}) \left(\frac{\partial N_L}{\partial N_S} \right)_{N_w}, \qquad (H.5)$$

where $\mu_{ij} = \frac{\partial^2 G}{\partial N_i \partial N_j}$ and $\mu_{LL} - 2\mu_{LH} + \mu_{HH} > 0$.

Thus, we can rewrite Eq. (H.4) as

$$\mu_S(T, P, N_L, N_H, N_S)$$

$$= \mu_S^{\text{fr}} - (\mu_{LL} - 2\mu_{LH} + \mu_{HH}) \left(\frac{\partial N_L}{\partial N_S} \right) \Delta N_L. \quad (H.6)$$

Note that $\left(\frac{\partial N_L}{\partial N_S} \right)_{N_w}$ is the effect of s on N_L at the point (T, P, N_L^0, N_H^0), i.e. the initial effect of an *infinitesimal* addition of dN_S to pure water. ΔN_L is the result of the addition of the *finite* quantity N_S.

We now assume that s stabilizes the component L, i.e. N_L increases from the initial state $(T, P, N_L^0, N_H^0, N_S)$ to the final state (T, P, N_L, N_H, N_S). Therefore, the second term on the RHS of Eq. (H.6) will be negative. Thus, upon relaxing the constraint on L and H, the chemical potential of s will decrease, which means that the structural changes induced by s will tend to *increase* the solubility. [For more details, see Ben-Naim (2009).]

The next situation is relevant to the solubility of s in a mixture of solvents: water w and a cosolvent, v.

We start with an initial state (T, P, N_w, N_S, N_v), add dN_v and examine the change in the chemical potential of s. Again, we view water as a mixture of L and H, and expand μ_S to first order in dN_v:

$$\mu_S(T, P, N_L + dN_L, N_H - dN_L, N_S, N_v + dN_v)$$

$$= \mu_S(T, P, N_L, N_H, N_S, N_v) + \left(\frac{\partial \mu_S}{\partial N_v}\right)_{N_S, N_L N_H}$$

$$+ \left(\frac{\partial \mu_S}{\partial N_L} - \frac{\partial \mu_S}{\partial N_H}\right) dN_L \tag{H.7}$$

or, equivalently,

$$\left(\frac{\partial \mu_S}{\partial N_v}\right)_{N_S, N_w} = \left(\frac{\partial \mu_S}{\partial N_v}\right)_{N_S, N_L, N_H}$$

$$+ \left(\frac{\partial (\mu_L - \mu_H)}{\partial N_S}\right)_{N_L N_H N_v} \left(\frac{\partial N_L}{\partial N_v}\right)_{N_S N_w}$$

$$= \left(\frac{\partial \mu_S}{\partial N_v}\right)_{N_S, N_L, N_H} - (\mu_{LL} - 2\mu_{LH} + \mu_{HH})$$

$$\times \left(\frac{\partial N_L}{\partial N_S}\right)_{N_w N_v} \left(\frac{\partial N_L}{\partial N_v}\right)_{N_w N_S}. \tag{H.8}$$

Note that in this case, we start from a state $(T, P, N_L, N_H, N_S, N_v)$ freeze-in the equilibrium $L \rightleftarrows H$ at N_L, N_H (not at pure water), and add dN_v. The effect of this addition on the chemical potential of s is

$$\Delta \mu_S = \left(\frac{\partial \mu_S}{\partial N_v}\right)_{N_S N_L N_H} dN_v - [\mu_{LL} - 2\mu_{LH} + \mu_{HH}]$$

$$\times \left(\frac{\partial N_L}{\partial N_S}\right)_{N_w N_v} \left(\frac{\partial N_L}{\partial N_v}\right)_{N_w N_S} dN_v. \tag{H.9}$$

We now distinguish between three possible cases:

(1) If both the solute s and the cosolvent v stabilize the L component, then the product of $\frac{\partial N_L}{\partial N_S}$ and $\frac{\partial N_L}{\partial N_v}$ is positive, and we get a negative contribution to the RHS of Eq. (H.9).
(2) If only one of the two cosolvents — either s or v — stabilizes L, then we get a positive contribution to Eq. (H.9).
(3) When both s and v destabilize L, we again get a negative contribution to the RHS of Eq. (H.9).

Although it is not possible to make any quantitative statements, I believe that these three cases correspond to the three slopes of the solubility (or the solvation Gibbs energy) of s in the mixtures of water and ethanol (see Figure 2.3).

It is probable that initially, when we add a small amount of alcohol to pure water, both s and v stabilize the structure of water (which we denoted L). The negative contribution to $\Delta\mu_S$ in Eq. (H.9) means that the solubility of s will initially increase. At larger concentrations of alcohol, s still stabilizes L, but the cosolvent v does not. This is probably the reason for the decrease in the solubility of s. At higher concentrations of v, the water is very diluted in v, there is not much of the L component left, and both s and v do not stabilize the L component.

It should be noted that the first term on the RHS of Eq. (H.9) is always negative. What we have examined above is the possible additional effects on $\Delta\mu_S$ due to structural changes induced by s and v.

Appendix I

Ben-Naim's "Pitfalls":
Don Quixote's Windmill

About a year after I published the article on "Levinthal's Question Revisited and Answered" [Ben-Naim (2012c)], Fang (2013) published an article with the title I use for this appendix. This is a very remarkable article. I wholeheartedly recommend reading it, not only to those who are interested in protein folding but also to anyone who is interested in science and in scientists. In this article, the author, who clearly did not understand anything in my article, has the chutzpah to criticize statements purportedly made by Ben-Naim, which in reality were never made or written by Ben-Naim.

I especially liked the amusing, ambiguous title. Did the author mean that Ben-Naim was Don Quixote, or only that the pitfalls which were discussed by Ben-Naim being compared to the windmills? I would appreciate it if someone could clarify the seeming parallelism posited by Fang.

318 | Myths and Verities in Protein Folding Theories

Here are examples of Don Fang's criticism of some imaginary windmills:

(1) In the abstract, the author states:

> *Ben-Naim, in three articles dismissed and "answered" the Levinthal paradox. . . .*

Indeed, I *dismissed* the Levinthal *paradox* in my articles, but I never "answered" the Levinthal paradox. I was careful enough to make it clear that I was answering the *Levinthal question*, and not the paradox.

(2) In the abstract, the author states:

> *He [Ben-Naim] claims no existence of Gibbs energy formula $G(X)$ where the variable is a protein's conformation X.*

By $G(X)$ the author presumably refers to the Gibbs energy landscape (GEL). This is absolutely untrue. In all my writings I discussed the function $G(X)$, and I never said that such a formula does not exist!

(3) In the abstract and in the article, the author claims:

> *His minimum distribution P_{eq} is wrong.*

This claim is surprising, considering that it comes from a mathematician. What I was talking about is the distribution P_{eq} which *minimizes* the Gibbs energy functional. I have no idea what the "minimum distribution P_{eq}" is, which the author claims to be mine. Again, this example proves that the author has no idea what is written in my articles which he is criticizing.

(4) In the introduction, the author writes:

> *Levinthal pointed out.... He then concluded that the natural protein folding must be caused-based, that is, the native structure has the (local) minimum of the Gibbs energy.*

Clearly, the author does not understand my distinction between "caused-based" and "target-based" theories of protein folding.

(5) In Section 1 of the article, the author writes:

> *Here Ben-Naim implies that the conformation of a protein should not be the variable of the Gibbs energy.*

Again, this is untrue. The so-called Gibbs energy landscape is a function of the conformation of the protein. I never claimed, or implied, that the "conformation of a protein should not be the variable of the Gibbs energy." I have no idea where the author got this absurd idea from.

(6) Another quotation from Section 1 of Fang's article:

> *Even knowing what is Ben-Naim's minimum distribution, we still do not know what the three-dimensional shape of the native structure.*

Again, I never talked about "Ben-Naim's minimum distribution." I have no idea what it is. As I noted above, what I wrote about is the distribution P_{eq} which minimizes the Gibbs energy functional, under a given environment (T, P, N). If one knew this distribution, then one could tell which conformations are more probable than the others under the given environment. Of course, I never claimed to know the 3D shape of the native protein.

(7) In Section 2, the author writes:

> *All previous attempts of deriving the Gibbs energy formula, including Ben-Naim's, missed the goal of identifying the three dimensional structure of a native protein.*

This quotation is doubly misleading. It is not only wrong but also quite amusing since, in a previous quotation above, the author claims that Ben-Naim "claims no existence of Gibbs energy formula. . . ."

If I claimed that the Gibbs energy formula does not exist (which I never did; see above), how then could I even attempt (among others) to derive the Gibbs free energy formula? In this quotation, the author contradicts his previous statements. The truth is that I never denied the existence of the Gibbs energy formula, and nor did I ever attempt to derive the Gibbs free energy formula. Therefore, all of the criticisms are addressed to claims that are either *inventions* by the author or figments of his imagination, because in reality no such claims appeared in my writings.

(8) In his conclusion, the author writes:

> *Ben-Naim's minimization at $P_{eq}(R)$ is analyzed and dismissed because it predicts that at equilibrium every possible conformation R will have the same probability.*

This comment proves beyond reasonable doubt that the author did not understand anything in my article. The parameter R in $P_{eq}(R)$ is the *location* of the *center* of a *simple spherical solute* in a solution. This is clearly stated in my article. The function $P_{eq}(R) = \frac{1}{V}$ is the equilibrium density function for the locations of the solute particle. It has nothing to do with proteins! Yet the author's criticism is that "Ben-Naim claims that $P_{eq}(R) = \frac{1}{V}$ for any conformation R," an absurd claim attributed to me but which I never made. The author failed to see that I was giving a simple example of a solute at point R, not a protein (see also Appendix J for this example).

There are many more misquotations and false statements in this article. I believe that the examples provided above suffice to demonstrate the fictitious character of the article.

I should end with a "positive" note on the author's achievements.

The author claims that he demonstrates "how to apply quantum statistics to derive the Gibbs energy formula $G(X)$, and the folding force $-\nabla G(X)$." Both of these are empty claims. All he did was to rewrite the Gibbs energy function $G(X)$ in terms of the entropy function $S(X)$, the volume function $V(X)$, and the energy function $U(X)$, but not an explicit Gibbs energy function $G(X)$. His achievement is summarized in Section 7.4 as:

Thus, by (27) we obtain the Gibbs free energy $G(X)$ . . .

$$G(X) = PV(X) + U(X) - TS(X)$$
$$= \sum_{i=1} \mu_i N_i(X) + \mu_c N_c(X).$$

A remarkable achievement indeed!

I therefore recommend reading (and enjoying) this totally irrelevant scientific article.

After submitting the manuscript of this book to the publisher, a new article by Fang appeared, entitled "Why Ben-Naim's Deepest Pitfall Does Not Exist" [Fang (2015)]. This article claims to show that what I claimed as a pitfall in Anfinsen's hypothesis [Ben-Naim (2012c)] is not a pitfall. Unfortunately, instead of checking his errors in his previous article [Fang (2013)], the author continues to invent statements, and then criticize them.

An example:

> *Thus, we may say that Ben-Naim actually falls into a pitfall: he thinks that only by considering an ensemble of conformations we can apply the second law of thermodynamics.*

I did not say that in my article. One can talk about the probability of finding a single conformation and also about the distribution of configurations. The Gibbs energy of one solute at a fixed location and conformation in a solvent is a well-defined system thermodynamically. This has been clarified many times in the past; see, for example, Ben-Naim (1992).

Then the author adds:

> *So the "deepest pitfall" of what Ben-Naim claimed comes from flawed inferences, misunderstanding of Anfinsen's thermodynamic hypothesis, and prejudice about how to apply the second law of thermodynamics.*

My claim is that most people *misinterpreted* Anfinsen's hypothesis. I have explained that in detail in Chapter 6. I have no idea what the author means by "prejudice" about the second law. Perhaps the author has "prejudice" — I don't!

In Section 2, the author writes:

> *If Ben-Naim noticed that there are actually two kinds of thermodynamic systems, single molecule and ensemble. . . he would not have claimed the deepest pitfall.*

This is really a silly comment! I did not claim that there are two kinds of thermodynamic systems. The author clearly confuses the thermodynamics of a dilute solution (in the limit of Henry's law) with the representation of the system by a system of pure solvent with one solute molecule. This is the same as representing an ideal gas by a single particle in a box. There are no two kinds of thermodynamic systems, and I never claimed that there are.

The article is full of absurdities. In particular, the author concludes that the quantity

$$\Delta G = k_B T \ln \rho \Lambda^3 < 0, \qquad (I.1)$$

which appears in my article, "*is fundamentally flawed.*" Why? "*Because with a little mathematics, it is inferred that*"

$$\lim_{\vec{R} \to \vec{R}_0} \inf G(T, P, N : \vec{R}) = -\infty. \qquad (I.2)$$

I have no idea from where he got this absurd result (with a little mathematics?). It certainly does not follow from my equation (I.1). [A similar absurdity can be found in his Eq. (4).] The meaning of the quantity defined in Eq. (I.1) has been explained many times in the past. It is the change in the Gibbs energy per particle when the particle is released from a fixed position, assuming classical statistical mechanics. It has nothing to do with what Fang calls the "Gibbs free energy function."

In conclusion, I still stand by everything I wrote in my article Ben-Naim (2012c), and in this book. I still recommend to anyone to read both of Fang's articles: Fang (2013, 2015).

They are very entertaining.

Appendix J

Examples of Gibbs Energy Landscapes and Gibbs Energy Functionals

In Chapter 6, we discussed the great confusion that is associated with Anfinsen's thermodynamic hypothesis. This confusion has inspired many to search for a "folding code," and many futile efforts have ensued to look for the "structure" of protein in the global minimum of the GEL.

To the best of my knowledge ever since Anfinsen enunciated his thermodynamic hypothesis, no one has asked himself or herself what is meant by the *minimum*, or the *global minimum*, which Anfinsen referred to as "the Gibbs free energy of the whole system is lowest."

As I have discussed in Chapter 6, Anfinsen's statement sounds as if he was referring to one formulation of the second law of thermodynamics. If that interpretation is accepted, then Anfinsen's hypothesis is trivial: an equilibrium state of a system, characterized by fixed values of T, P, N, must be at a Gibbs energy minimum.

Unfortunately, most researchers in the field of pro-tein folding (including mathematicians; see Appendix I) misinterpreted Anfinsen's hypothesis, and concluded that the conformation of the native structure of the protein must be at the global minimum of the GEL. This interpretation, though very common, is unwarranted.

This appendix is devoted to clarifying the difference between the two possible "minima of the Gibbs energy." We will do this by a few examples.

(a) A Two-Compartment System

Consider a system of two compartments, as shown in Figure J.1. Initially, the two compartments are characterized by the variables N_1, V, T and N_2, V, T, respectively, in Figure J.1. If the system as a whole is isolated, then upon removing the constraints on the motion of particles from one compartment to another (i.e. opening the barrier between the two compartments; [Figure J.1(b)], the entropy change must be positive:

$$dS = \frac{\partial S}{\partial N_1}dN_1 + \frac{\partial S}{\partial N_2}dN_2 = \frac{(\mu_1 - \mu_2)dN_1}{T} > 0.$$

Thus, if $\mu_1 - \mu_2 > 0$, then $dN_1 > 0$ and if $\mu_1 - \mu_2 < 0$, then $dN_1 < 0$.

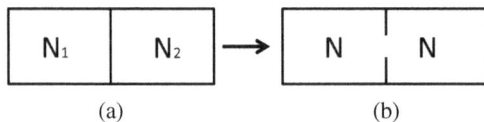

(a) (b)

Fig. J.1. The process of equalization of the density in the entire system.

The second law of thermodynamics states that the entropy *function* $S(E_T, V_T, N_T; x_1)$ has a single maximum, i.e.

$$\left(\frac{\partial S}{\partial x_1}\right)_{E_T, V_T, N_T} \tag{J.1}$$

where E_T is the total energy of the system, $V_T = 2V$ is the total volume of the system, and $N_T = N_1 + N_2$ is the total number of particles. x_1 is defined by

$$x_1 = \frac{N_1}{N_1 + N_2}. \tag{J.2}$$

In this particular example, we have only one variable, x_1, with respect to which the entropy has a maximum (E_T, V_T, N_T constant).

If the same system is at a constant temperature T, then the Helmholtz energy has a *single minimum* at equilibrium, i.e.

$$\left(\frac{\partial A}{\partial x_1}\right)_{T, V_T, N_T} = 0. \tag{J.3}$$

Before we generalize, we note that the parameter x_1 is also the distribution (x_1, x_2), with $x_1 + x_2 = 1$.

(b) Generalization to *n* Compartments

The generalization of the previous example is shown in Figure J.2. We start with an isolated system with an arbitrary

Fig. J.2.   Generalization of the process in Fig. J.1.

initial distribution of particles in n compartments, $x = x_1, \ldots, x_n$, where $x_i = N_i/N$ is the mole fraction of particles in compartment i. If we release the constraint on constants N_i, i.e. we let particles move freely between the compartments, the entropy of the system must increase. At equilibrium, the entropy will attain a maximum with respect to all possible *distributions* x, i.e. there is a *single maximum* of $S(E_T, V_T, N_T; x)$ with respect to x, at constant E_T, V_T, N_T. One can easily generalize to the continuous limit and prove that for an isolated system the entropy has a maximum with respect to the density distribution $\rho(R)$, which is the local density of particles in a small element of volume $dxdydz$ in the system. At equilibrium, there is a density $\rho^{(eq)}R = N_T/V_T$ which maximizes the entropy functional $S[E_T, V_T, N_T; \rho(R)]$, and this density is the uniform density. [For details, see Ben-Naim (2008, 2011d)].

We note also that one can get the Boltzmann density distribution of particles in a gravitational field by maximizing the entropy functional of a column of air in a gravitational field [see Ben-Naim (2008)].

The point we emphasize here is that, in general, the second law when applied to an isolated system states that the maximum entropy (or the minimum Gibbs or Helmholtz energy, in the case of T, V, N) is obtained for a *specific distribution function*. This means that the entropy (or the Helmholtz or Gibbs energy) is viewed as a *functional* of this distribution.

The next few examples apply to molecular distributions.

(c) A Solute at a Fixed Position R_0

Consider an isolated system having a fixed energy E, volume V and N solvent particles. We place a solute s at a fixed location R_0. We will refer to this particle as the *solute*. For simplicity, we assume that this solute differs from the solvent molecules (Figure J.3). Also for simplicity, we look at one solute in a macroscopic system of N solvent molecules.

For such a system, one can define the entropy function $S(E, V, N; R_0)$. Clearly, by fixing the location of the solute in the system, we have a constrained equilibrium state. Here, the parameter R_0 is the microscopic parameter characterizing the constrained equilibrium state.

What happens if we remove the constraint and let the solute particle wander freely in the system?

Here is a potential pitfall. We write the entropy *function* as $S(E, V, N; R)$, and we keep E, V, N constant. Since E, V, N

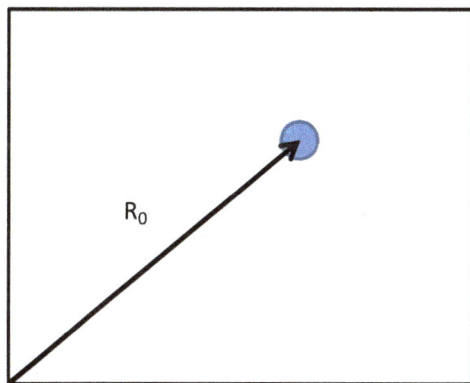

Fig. J.3. A solute at a fixed position in a solvent.

are constants, the entropy is a function of one variable, R. If we release the constraint on R_0, we are inclined to conclude that the entropy, as a *function* of R, must increase toward a maximum.

Of course, no one has reached that conclusion for this particular example. But many have reached essentially the same conclusion for more complicated cases, which are discussed below. The reason we naturally avoid falling into this pitfall is that in this particular case, even without any calculations, we feel that when the location of the particle is not constrained at R_0, the entropy is not a monotonically increasing function of R, and there exists no specific location — say, R_{max} for which the entropy is maximal. In fact, it is easy to prove using Shannon's theorem that at equilibrium (after lifting the constraint on R) the probability of finding the particle at any point R within the system is constant, independently of R. Specifically, the probability density is

$$P_{eq} \to P_{eq}(R) = \frac{1}{V} \quad \text{for any } R. \tag{J.4}$$

Thus, we are still left with the question posed earlier: Maximum with respect to what?

At this stage, we have two different "functions." One is $S(E, V, N; R)$, and the other is a functional $S[E, V, N; P(R)]$. The second law states that the entropy of the system (having a fixed value of E, V, N) is maximum with respect to all possible *distributions of locations.*

Let $P(R)$ be the distribution of locations of the particle, such that

$$\int_V P(R)dR = 1, \tag{J.5}$$

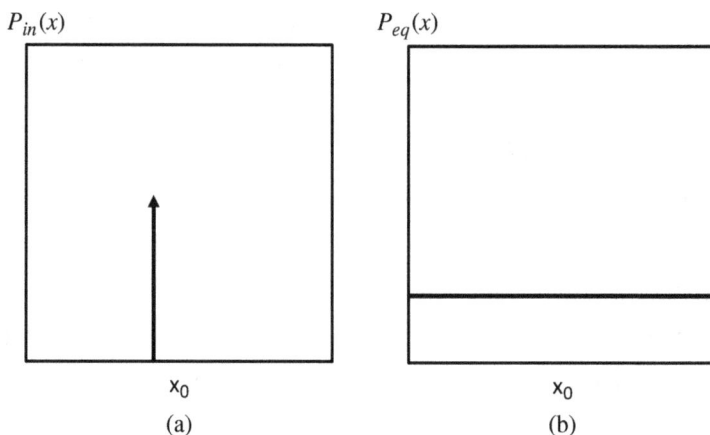

Fig. J.4. The initial and the final distribution of the location of the solute in one-dimensional system.

i.e. the probability of finding the solute particle at *any* location within the volume V is 1. Instead of the function $S(E, V, N; R)$ we look at the *functional* $S(E, V, N; P(R))$. Shannon's theorem states that the entropy functional will be maximum with respect to the *function* $P(R)$. Thus, in the initial state, the entropy functional is written as $S(E, V, N; \delta(R - R_0))$, where $\delta(R - R_0)$ is the Dirac delta function (Figure J.4). At equilibrium, there exists a *single function* [which we denoted $P_{eq}(R)$] that maximizes the entropy, i.e. the condition for the maximum is that the functional derivative is zero:

$$\frac{\delta S(P(R))}{\delta P(R)}\bigg|_{E, V, N} = 0. \qquad (J.6)$$

The distribution function that maximizes the entropy functional is given in Eq. (J.4) and for a 1D system, it is shown in [Figure J.4(b)].

Before we turn to the next examples, we reformulate the same principle in terms of the Gibbs energy. For a system characterized by the variables T, P, N, we can write the Gibbs energy *function* of the system as $G(T, P, N; R)$.

Note that $G(T, P, N; R)$ is not a monotonically decreasing function of R and that there exists no value of R for which G is minimal. Instead, the *functional* $G(T, P, N; P(R))$ has a single minimum with respect to all possible distributions $P(R)$. The distribution $P_{eq}(R)$, for which G is minimal, is given in Eq. (J.4).

In this example, we found that G as a function of R is constant. In the next examples, we will encounter a Gibbs energy function that has one or more minima. The point to stress here is that these minima, when they occur, are not a result of the second law. The second law requires that the entropy *functional* $S(E, V, N; P(R))$ or the Gibbs energy *functional* $G(T, P, N; P(R))$ should have a *single* extremum with respect to the *probability distribution* $P(R)$. The equilibrium distribution $P(R)$ may have any form, with or without minima and maxima.

(d) Two Particles in the Solution

Suppose that we start with N solvent molecules at a given temperature T and pressure P, with the two solute particles at a fixed locations R_1 and R_2, and at a distance r_0 from each other [Figure J.5(a)]. The system is at equilibrium in this initial state. Now, we release the constraint imposed on the *fixed* location of one of the particles — say, the one at R_1 — leaving the second particle at a fixed position R_2.

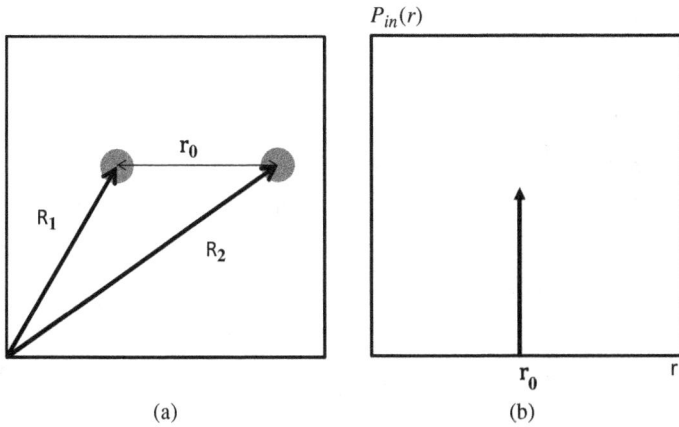

Fig. J.5. The initial state and initial distribution of two particles as a distance r.

Thermodynamically, because the released particle is no longer confined to a fixed location, the Gibbs energy of the system will be reduced by the amount

$$\Delta G = k_B T \ln \rho \Lambda^3 < 0. \qquad (J.7)$$

This is the same as the change in the Gibbs energy in the previous example.

Thus, the *system* as a whole reaches a new minimum for two reasons: the release of the constraint on a fixed position, which is ΔG in Eq. (J.7) and the change in the distance distribution (see below).

Here is the potential pitfall. The GEL, for this example, is the *function* $G(T, P, N; r)$, where $r = |R_1 - R_2|$. This function (note that T, P, N are constants) has several minima and maxima. When the constraint on the fixed distance r is released, the system will *not fall* into one of the minima of the GEL. In other words, thermodynamics does not require that the released particle move toward the minimum of the *function* $(T, P, N; r)$. In our example, this function has several minima. Instead, the system will proceed from the initial state, fixed r_0, to a new state, having a *new distribution* of distances — not a new "configuration" r. The second law requires that the *functional* $G(T, P, N; P(r))$ should have minimum with respect to all possible distributions $P(r)$. In the initial state, the Gibbs energy *functional* was $G(T, P, N: \delta(r - r_0))$, where $\delta(r - r_0)$ is the Dirac delta function [Figure J.5(b)]. In the final state, the Gibbs energy functional is $G(T, P, N: P_{eq}(r))$, where $P_{eq}(r)$ is related to the PMF between the two solutes. Thus, upon releasing the constraint on fixed r, the system will not proceed to a new minimum in the Gibbs energy *function* $G(T, P, N; r)$, but to a new minimum in the Gibbs energy *functional* $G(T, P, N; P(r))$. The GEL may or may not have a minimum. The Gibbs energy *functional* must have a *single minimum* with respect to the distribution function $P(r)$. (See also Appendix K.)

Figure J.6 shows two possible GEL, for the system of two solute particles in a solution. $W(r)$ is the Gibbs energy change for the process of bringing the two particles from infinite separation to the distance r, in the liquid, keeping T, P, N constant. [Figure J.6(a)] is the PMF for the case of very dilute solvent; in this case, the PMF reduces to the pair potential $U(r)$. [Figure J.6(b)] shows the PMF for a higher density of the solvent. Here, we see that there are a few minima and maxima. The probability distribution is related to the solute–solute pair correlation function $g_{ss}(r)$ (See Figure J.7).

$$P(r)dr = \rho_s^2 g_{ss}(r)4\pi r^2 dr \qquad \text{(J.8)}$$

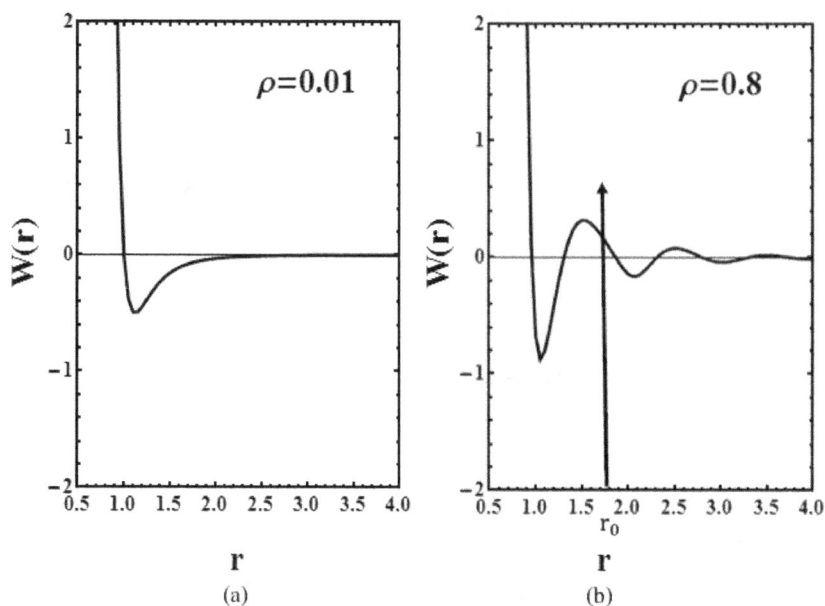

Fig. J.6. The PMF or the GEL of the two solute in solution at two-solvent densities.

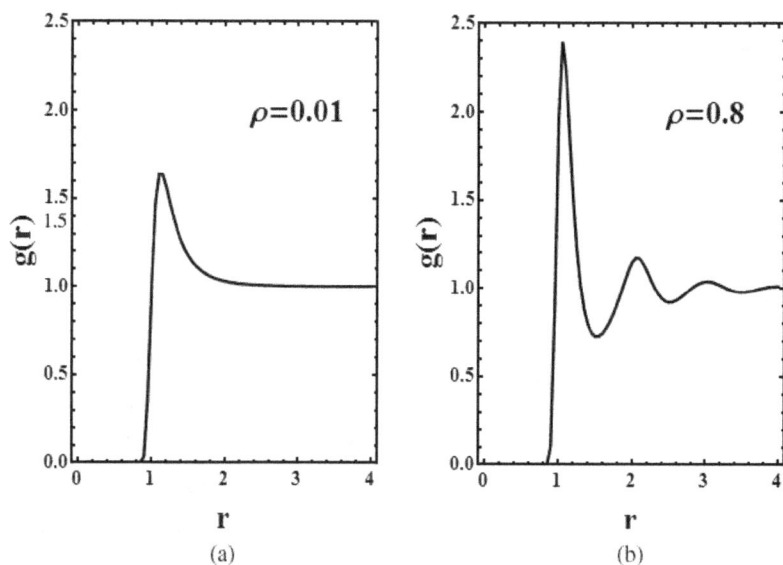

Fig. J.7. The pair correlation function for the two solutes in a solvent at two-solvent densities.

and the pair correlation function is related to the PMF by

$$W(r) = -k_B T \ln g_{ss}(r). \tag{J.9}$$

Note that for two solutes HS, the pair potential, i.e. the limit of $W(r)$ when $\rho \to 0$ has no minima or maxima.

(e) A Solute with One Internal Rotational Degree of Freedom

The next example brings us one-step closer to proteins. Consider again a system at T, P, N and N_a solute molecules. We assume that the solutes are very dilute in the system, i.e. $N_a \ll N$, so that we can neglect solute–solute interactions. We start with N_a ethane molecules, each at a fixed positions R_i, and all at a fixed dihedral angle ϕ [Figure J.8(a)]. If we release the constraints on fixed positions and fixed angle ϕ_0,

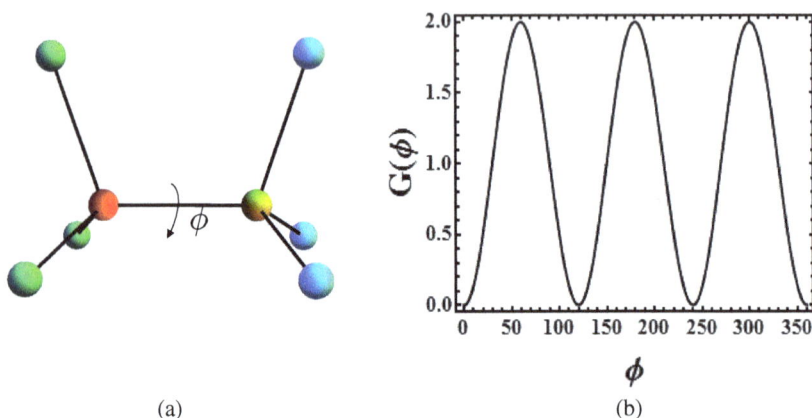

(a)   (b)

Fig. J.8. The general form of the GEL of ethane.

the system's Gibbs energy will be reduced, for three reasons: (a) the acquired kinetic energy of translation, which is essentially N_a times the quantity (J.7), but with $\rho = N_a/V$, the density of the solute, (b) the acquisition of the rotational degree of freedom, and (c) the change in the potential energy of the molecule. We are interested only in the latter. Thus, we can think of N_a solutes at fixed positions R_i, and far from each other, each at a fixed angle ϕ. We release the constraint only on the fixed angle of internal rotation. The initial probability distribution of angles is

$$P_{\text{in}}(\phi) = \delta(\phi - \phi_0), \tag{J.10}$$

where $\delta(\phi - \phi_0)$ is the Dirac delta function. After the release of the constraint on a fixed angle, there will be some distribution of angles, $P_{\text{eq}}(\phi)$. The GEL, for this example, is a function of the form shown in [Figure J.8(b)]. The change in the Gibbs energy upon releasing the constraint on a fixed ϕ is

$$\Delta G = G[T, P, N, N_a; P_{\text{eq}}(\phi)]$$
$$- G[T, P, N, N_a; \delta(\phi - \phi_0)] \leq 0. \tag{J.11}$$

When we release the constraint on a fixed ϕ, the system will *not* move to a new angle ϕ for which the GEL has a minimum (this function has three minima and three maxima; [Figure J.8(b)], but to a new distribution of angles $P_{\text{eq}}(\phi)$ for which the Gibbs energy functional has a single minimum at $G[T, P, N, N_a; P_{\text{eq}}(\phi)]$. This minimum was shown schematically in Figure 6.3.

Next, we take the ethane molecules and substitute some of the hydrogen atoms with LJ spheres of different diameters

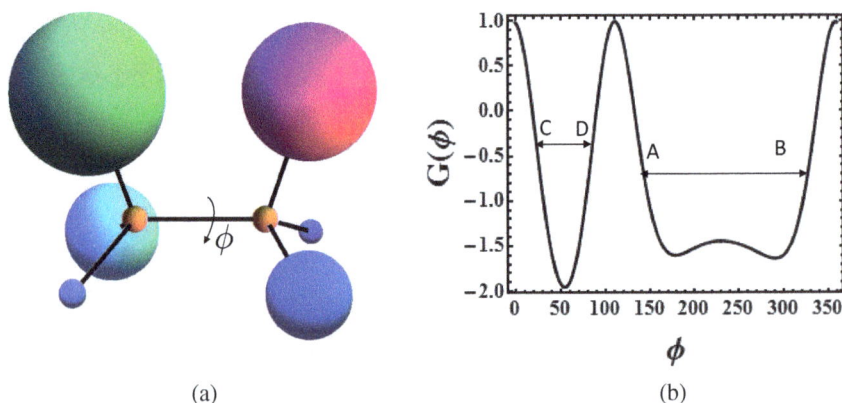

Fig. J.9. The GEL of a substituted ethane molecule.

σ and energy parameter ε [Figure J.9(a)]. The details of the calculation are not important; we can think of a whole protein at a fixed conformation, and we allow rotation only about one bond. The GEL for this case might look like the function shown in [Figure J.9(b)].

As in the previous example, if we start with a fixed angle ϕ_0 and release this constraint, the system will proceed to a new distribution, such that the Gibbs energy *functional* has a minimum.

In all the examples given above, if there exists a configuration for which the probability is relatively large, we can say that this state is a stable state, and in this stable state, the GEL has a minimum. In addition, the Gibbs energy functional must always have a minimum with respect to all possible distributions of the states.

There is one more important aspect to the distribution function $P(\phi)$ that should be taken into account. In all the examples above, we *calculated* the Gibbs energy function $G(T, P, N; \phi)$. We then converted from this function to probability distribution. This conversion is valid provided that all the configurations, such as all ϕ's in [Figure J.9(b)], are *accessible* in the "environment" T, P, N, where we have released the constraint on ϕ. This is a very important point to consider before we convert energies, or Gibbs energies into probabilities.

Suppose that whenever we synthesize the molecule, we get an initial conformation which is in the region denoted AB in [Figure J.9(b)]. If the potential barrier to crossing from AB to the other regions of the configurational space is very large, then the entire configurational space is not $0 \leq \phi \leq 2\pi$, but only the accessible region within AB. In this case, the conversion from the GEL to probabilities is

$$P(\phi) = \frac{\exp[-\beta G(T, P, N; \phi)]}{\int_{AB} \exp[-\beta G(T, P, N; \phi)]d\phi}. \qquad (J.12)$$

Note that the normalization constant is an integral over all *accessible* states, not over the entire range of configurations, $0 \leq \phi \leq 2\pi$. Furthermore, the range of accessible states AB will depend on the temperature, pressure and solvent composition. This means that the conversion from $G(T, P, N; \phi)$ to $P(\phi)$ is dependent on T, P, N.

In the particular example shown in [Figure J.9(b)], there is a finite probability of finding the system in any small interval $d\phi$ within the region AB, but there is a zero probability of finding the molecule in any configuration outside AB. The normalization condition for the distribution function $P(\phi)$, in this case, is

$$\int_{AB} P(\phi)d\phi = 1. \qquad (J.13)$$

This is an important aspect of the probability distribution, or of the GEL, which is easy to grasp in the simple example discussed above. It means that the only region of the GEL which is of interest is the region AB. There is no need to study the *entire* GEL, simply because there are many regions that are inaccessible to the system at a given T, P, N.

There is one more important aspect of Figure J.9 besides the region of accessibility in the GEL, and it is that there might be minima in the GEL that are *lower* than the minima in the accessible region. For instance, if the accessible region is AB in [Figure J.9(b)], then the minimum at about $\phi = 50°$ is lower, but not accessible. Thus, the search for a global minimum in the entire GEL is not always relevant to the most stable state within the accessible part of the GEL.

If, on the other hand, our initial angle is in the region CD, then the accessible states are in the region CD in Figure J.9(b). In this case, the probability distribution is [Figure J.9(b)] not (J.12) but

$$P(\phi) = \frac{\exp[-\beta G(T, P, N; \phi)]}{\int_{CD} \exp[-\beta G(T, P, N; \phi)]d\phi}, \qquad (J.14)$$

with the normalization condition

$$\int_{CD} P(\phi)d\phi = 1. \qquad (J.15)$$

In this case, the accessible region is much smaller than the entire configurational space. Clearly, it would be a waste of time to study the entire configurational space.

The abovementioned point should also be taken into account when one is performing simulation of protein folding. On paper, or rather *in silico*, we can start the simulation from any arbitrarily chosen configuration and follow its trajectory with time. The final stable configuration which the protein will reach, if it reaches such a state, depends on the region of states which are accessible from the initial states. For instance, in Figure J.9, if we start from a value of ϕ in the region CD, then at equilibrium most of the molecules will be at the minimum at the point $\phi = 50°$. We can say that this is the most stable state. However, if we start from an initial configuration in the region AB, the system will not be settled in any particular configuration. Note also that in this case, the system's configurations will occupy the region AB, although the minimum at $\phi = 50°$ is deeper than any of the minima in AB.

(f) A Solute with Two Internal Rotational Degrees of Freedom

In the previous example, we had one internal angle of rotation. This allowed us to draw the GEL of the molecule as a one-dimensional function $G(\phi)$. The next example is a molecule with two rotational angles. One way to arrive at the new model is to start with a protein at a fixed location and conformation in the liquid at some fixed T, P, N. We now release the constraint imposed on the conformation by allowing rotations only about *two bonds* — say, ϕ and ψ — in Figure J.10. All other rotational angles remain "frozen-in." We model this case by taking a propane molecule, (Figure 6.16), and replace some of the hydrogen atoms by Lennard–Jones spheres. As we have done in the previous

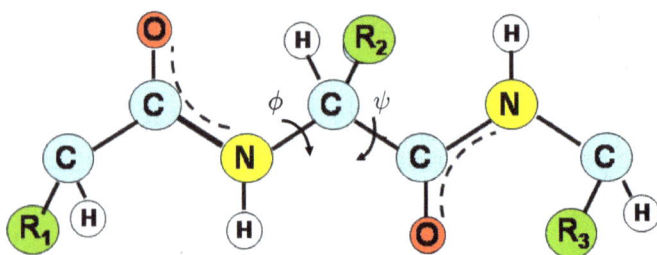

Fig. J.10. A segment of a protein with two possible rotational degrees of freedom.

example, we can calculate the EL, and the GEL, $G(\phi, \psi)$, for this case. The details of the model and the calculations are not important. The resulting GEL is a two-parameter function and may have a relatively simple form with a few minima, as in Figure 7.3, or may have a very complicated form, as in Figure 6.16. Again, the details are not important. Just have a look at Figure 6.16 and be impressed by the following: first, there are many minima and maxima, second, there exists no single outstanding minimum and, third, if the molecule is synthesized (or the initial configuration in a simulation) at, or in the vicinity of, a minimum, then the protein can access only a tiny region in the GEL from this configuration. There remains a vast region in the GEL which is not accessible.

These features should be borne in mind whenever one discusses the GEL of a real protein. Of course, in this case we cannot visualize the form of the GEL, but we can imagine its extreme complexity. The GEL of a protein is a function of at least $2M$ angles, where M is of the order of a few hundreds.

(g) The Case of Real Protein

Suppose we start with a protein at a fixed location and conformation in the solvent. We already know from previous examples that if we release only the constraint on a fixed position the change in the Gibbs energy will be $k_B T \ln \rho_p \Lambda_p^3$. If we also release the constraint on the fixed conformation, the Gibbs energy will further decrease, for two reasons, the kinetic rotational energy, and the change in the PMFs and torques associated with the rotations about the chemical bonds. If we are interested in the GEL, we can imagine releasing the constraint on each fixed angle consecutively, i.e. we select one bond about which we allow the protein to rotate, and all other internal rotation angles are held fixed. The form of the GEL for this case is extremely complicated. We can proceed by removing the constraint on one angle ϕ_i at a time. Eventually, we will see that the GEL function is $G(T, P, N; \phi_1, \Psi_1, \ldots, \phi_M, \Psi_M)$, where the configuration of the entire protein is given by the $2M$ dihedral angles $\phi_1, \Psi_1, \ldots, \phi_M, \Psi_M$. Starting from any initial configuration $\phi_1^{in}, \ldots, \Psi_M^{in}$ (say, of the unfolded form), the Gibbs energy function has the value $G(T, P, N; \phi_1^{in}, \ldots, \Psi_M^{in})$. When we release the constraint of a fixed configuration, the Gibbs energy of the system will reach a new minimum; it is not a minimum with respect to the variables $\phi_1, \ldots, \Psi_M$, but with respect to all possible distributions $P(\phi_1, \ldots, \Psi_M)$, i.e. the Gibbs energy *functional* has a minimum at some distribution $P_{eq}(\phi_1, \ldots, \Psi_M)$, with the value of $G(T, P, N; P_{eq}(\phi_1, \ldots, \Psi_M))$.

The GEL will have many minima and maxima. Some regions of the GEL will be completely inaccessible, and many regions will be disconnected, separated by a high (perhaps infinite) barrier. The portrayal of such a GEL as a funnel, or as any other shape having one outstanding minimum is totally unrealistic. The native structure of the protein, if such a structure exists, must be located at some minimum in the GEL, but for large proteins this minimum is unlikely to be the global minimum of the entire protein.

Besides the irrelevance of the funnel model for protein, there are a few amusing aspects of the representation of the funnel shape in a 3D diagram. The GEL is supposed to describe the Gibbs energy as a function of the coordinates ϕ and ψ. In a 3D diagram, the x, and y, axes are supposed to represent ϕ and ψ, while the z, axis represents the Gibbs energy of the system. Curiously, the span of the x, and y, axes is sometimes denoted "entropy," and the y, axis "energy." This is quite strange. If by "energy" one strictly means the *energy* of the protein, then the whole figure is irrelevant to proteins in solutions. On the other hand, if the z, axis represents the Gibbs energy of the system, then it includes both the energy and the entropy, and one does not need to specify the entropy on a separate axis or axes. Sometimes, it is commented that the inclusion of the entropy in the funnel picture is only to represent the entire configurational space of the unfolded form compared with the smaller region of the folded form. This explanation is also misleading for two reasons: (a) the entropy is not represented by the entire configurational space, but only by that part of it which is accessible to the unfolded form, a part which we know nothing about except that it is much larger than the part representing the folded form, and (b) even if one focuses on the accessible states of the unfolded form, the entropy is a measure of the *size* of that configuration space only when all of the accessible states have equal energy, which is clearly not the case for the unfolded protein.

Appendix K

The Existence of a *Single* Minimum of the Gibbs Energy Functional

Note the italicization of the word "*single*" in the title of this appendix.

If you are going to apply the second law of thermodynamics to the protein folding problem, remember that there is no need to find out of the myriad possible minima, a specific minimum in the GEL. The second law states that there is a *single minimum* for a system at equilibrium. This is what we are going to prove in this appendix.

In this appendix, we consider a system of N_s solute molecules (protein) in a solvent consisting of N_w solvent molecules (water) at a given temperature T and pressure P. We will assume that the solute is much diluted in the solvent ($N_s \ll N_w$). If it consists of several components, then we reinterpret N_w to be the composition vector of all solvent molecules. The triplet of variables T, P, N_w will be referred to as the "environment."

Each solute molecule is characterized by its conformation, which is determined by specifying all the angles of internal rotation. For simplicity of notation, we assume that there is only one angle of internal rotation, denoted ϕ, as in Figure A.2. Initially, we assume that there are a finite number of conformations characterized by the angles $\phi_1, \ldots, \phi_m$. Later, we take the limit of a continuous variation in ϕ.

Note carefully that in this appendix, we characterize the *conformation* of a single solute by *one* rotational angle, attaining m different (discrete) values, $\phi_1, \ldots, \phi_m$ (as in Figure A.2), with $m = 4$. This is different from the notation in the previous appendix J, where the conformation of the protein is characterized by n different rotational angles $\psi_1, \ldots, \psi_n$, each of which can attain m different values.

Thus, our system consists of N_s solute molecules distributed in m different conformations, $\phi_1, \ldots, \phi_m$. We denote by $N_1, \ldots, N_m$ the *composition* of the solute molecules, where N_i is the average number of solute molecules in the conformation ϕ_i. We also assume that all of the m conformations are *accessible* in the given *environment* T, P, N_w.

The system can be viewed in two equivalent ways. The first is as a two-component system, w and s. In this view, the Gibbs energy of the system is given by

$$G = N_w \mu_w + N_s \mu_s. \qquad (K.1)$$

In the second way, we view the system as $m + 1$ components, with the composition $N_w, N_1, \ldots, N_m$. In this view, we express the Gibbs energy as

$$G = N_w \mu_w + \sum_{i=1}^{m} N_i \mu_s(i), \qquad \text{(K.2)}$$

where $\mu_s(i)$ is the chemical potential of the solute being at a specific conformation, ϕ_i. This view may be referred to as the *mixture model* (MM) view applied to the solute molecules.

At equilibrium the Gibbs energy must have a minimum with respect to the variables $N_1 N_2, \ldots, N_m$ with the constraint $\sum_{i=1}^{m} N_i = N_s$. This is a statement of the second law of thermodynamics.

The condition for the minimum is easily obtained by using the method of Lagrange multipliers. We define the auxiliary function:

$$K(N_1, N_2, \ldots, N_m) = G(N_1, \ldots, N_m)$$
$$- \lambda \left(\sum_{i=1}^{m} N_i - N_s \right). \qquad \text{(K.3)}$$

We take the derivatives with respect to all N_i, $i = 1, \ldots, m$:

$$\frac{\partial K}{\partial N_i} = \frac{\partial G}{\partial N_i} - \lambda = \mu_s(i) - \lambda = 0. \qquad \text{(K.4)}$$

Hence

$$\mu_s(i) = \lambda, \quad \text{for each } i. \qquad \text{(K.5)}$$

At equilibrium, we must have the equality

$$G = N_w \mu_w + N_s \mu_s = N_w \mu_w + \sum_{i=1}^{m} N_i \mu_s(i). \qquad \text{(K.6)}$$

This means that the Gibbs energy of the system is unchanged if we choose to view it as a mixture of "different" solutes at equilibrium. The justification for the use of the MM approach to any liquid, and in particular to water, is discussed in Ben-Naim (2009). From Eqs. (K.4) and (K.5), we have the equality

$$\mu_s = \mu_s(i) = \mu_s(2) = \cdots \mu_s(m), \qquad (K.7)$$

which is simply the condition of equilibrium between all the m species.

We define the mole fractions of the species i by

$$x_i = \frac{N_i}{N_s}. \qquad (K.8)$$

So far, we have used the second law to claim that the *function* $G(T, P, N_w; x_1, \ldots, x_m)$ has a minimum at equilibrium, with respect to the variables $x_1, \ldots, x_m$, while keeping the variables P, T, N_w constant.

Equivalently, we can say that the function $G(T, P, N_w; x_1, \ldots, x_m)$ has a minimum at equilibrium with the constraint $\sum_{i=1}^{m} x_i = 1$. Note carefully that $x_1, \ldots, x_m$ describes the *composition* of the solute, not the *conformations* of the solute as in the GEL.

Next, we show that this minimum is a unique one, i.e. that there exists a unique distribution of species $x_1, \ldots, x_m$ which is referred to as the *equilibrium distribution* at which the Gibbs energy has a single minimum. We will also find out the equilibrium distribution of species.

To do this, suppose that we start with any arbitrary initial distribution of species, denoted $x_1^{in}, x_2^{in}, \ldots, x_m^{in}$. We will show that starting with *any* arbitrary initial distribution $x_1^{\text{in}}, \ldots, x_m^{\text{in}}$, when releasing the constraint on this distribution, the system will evolve to a unique distribution, $x_1^{\text{eq}}, \ldots, x_m^{\text{eq}}$, for which the Gibbs energy is at a minimum, i.e. we will show that the Gibbs energy change for the process

$$\left(x_1^{\text{in}}, \ldots, x_m^{\text{in}}\right) \rightarrow \left(x_1^{\text{eq}}, \ldots, x_m^{\text{eq}}\right) \qquad \text{(K.9)}$$

is *always* negative for any initial distribution. This is true provided that all the states, i.e. the angles $\phi_1, \ldots, \phi_m$, are accessible. For example, if we start with, say, $x_1 = 1$ and all $x_i = 0$, $i \neq 1$, the system will relax to the final equilibrium composition as in (K.9).

We write the chemical potential of the solute s as

$$\mu_s = \mu_s^* + k_B T \ln \rho_s \Lambda_s^3, \qquad \text{(K.10)}$$

where Λ_s^3 is the momentum PF of the solute s, $\rho_s = N_s/V$ is the number density of s, and μ_s^* is the pseudochemical potential defined in Eq. (K.10). For very dilute solutions, μ_s^* is independent of ρ_s. The pseudochemical potential is the Gibbs energy change for inserting a solute at a *fixed* position in the solvent. [For details, see Ben-Naim (2006).]

Viewing the solute as a mixture of species, we can write the chemical potential of each species as

$$\mu_s(i) = \mu_s^*(i) + k_B T \ln \rho_i \Lambda_i^3, \qquad \text{(K.11)}$$

where $\rho_i = N_i/V$ and $\Lambda_i^3 = \Lambda_s^3$ for all i. We have denoted by $\mu_s(i)$ the chemical potential of the solute at a specific conformation ϕ_i.

From Eqs. (K.10) and (K.11), and the condition of equilibrium (K.8), we get the equilibrium distribution of the species:

$$x_i^{eq} = \frac{\rho_i}{\rho_s} = \exp\left[\beta\left(\mu_s^* - \mu_s^*(i)\right)\right]. \qquad (K.12)$$

Summing over all the species i, we get from Eq. (K.12)

$$1 = \sum_i x_i^{eq}$$

$$= \exp\left(\beta\mu_s^*\right) \sum_{i=1}^{m} \exp\left[-\beta\mu_s^*(i)\right]. \qquad (K.13)$$

This can be rearranged to

$$\exp\left(-\beta\mu_s^*\right) = \sum_{i=1}^{m} \exp\left[-\beta\mu_s^*(i)\right], \qquad (K.14)$$

which is the general relationship between the pseudochemical potential of the solute s and the pseudochemical potentials of all of the solute species.

Since each of the terms on the RHS of Eq. (K.14) is positive, we must have the inequality

$$\exp\left(-\beta\mu_s^*\right) \geq \exp\left[-\beta\mu_s^*(i)\right], \quad \text{for each } i, \qquad (K.15)$$

or, equivalently,

$$\mu_s^* \leq \mu_s^*(i), \quad \text{for each } i. \qquad (K.16)$$

The last inequality means that if we start with a system of N_s solute molecules, each at a *fixed position*, and *fixed conformation* ϕ_i, releasing the constraint on a fixed conformation will always result in *lowering* the Gibbs energy of the system, i.e.

$$N_s \mu_s^* - N_s \mu_s^*(i) \leq 0, \quad \text{for each } i. \tag{K.17}$$

We now generalize these results in two steps. First, suppose that we start with N_s solute molecules, each at a fixed position (but they are far from each other) with the composition $x_1^{in}, x_2^{in}, \ldots, x_m^{in}$, and relax the constraint on the fixed composition. The Gibbs energy change of the system must be negative.

To show that, we multiply (K.17) by x_i^{in} and sum over i to obtain

$$\sum_{i=1}^{m} x_i^{in} \left[\mu_s^* - \mu_s^*(i) \right] = \mu_s^* - \sum x_i^{in} \mu_s^*(i)$$

$$= \sum_{i=1}^{m} x_i^{eq} \mu_s^* - \sum x_i^{in} \mu_s^*(i) \leq 0. \tag{K.18}$$

This is an important inequality. It states that the change in the Gibbs energy of a system of solute particles at fixed positions (but far from each other) from an initial distribution x^{in} to the final distribution x^{eq} is always negative. This quantity may be referred to as the *intrinsic* change in the Gibbs energy, due to the release of the constraint only on a fixed distribution of conformations.

It should be noted that an inequality of the type (K.16) does not hold in general for the chemical potentials of s and of the species i. In general, we have

$$\mu_s - \mu_s(i) = \mu_s^* - \mu_s^*(i) + k_B T \ln \rho_s \Lambda_s^3 - k_B T \ln \rho_i \Lambda_i^3$$
$$= \left[\mu_s^* - \mu_s^*(i)\right] + \left[-k_B T \ln x_i\right]. \qquad (K.19)$$

The first term in square brackets on the RHS of (K.19) is negative by (K.16). The second term is always positive. Therefore, the sign of the difference $\mu_s - \mu_s(i)$ depends on whether x_i is larger or smaller than x_i^{eq} [Eq. (K.12)]. On the other hand, the *average* of $\mu_s - \mu_s(i)$ with the distribution x^{in} is always negative; see Eq. (K.18).

Next, we prove a stronger statement. We start with any initial composition of the solute molecules $x^{in} = \left(x_1^{in}, \ldots, x_m^{in}\right)$. The solute particles are not restricted to fixed positions. They are free to wander in the system. We relax the constraint on the fixed conformations of the species and let the system relax to a new equilibrium state. We will show that the Gibbs energy change in this process must be negative.

The total change in the Gibbs energy for the process $x^{in} \to x^{eq}$ is

$$\Delta G = [N_w \mu_w + N_s \mu_s] - \left[N_w \mu_w + \sum_{i=1}^{m} N_i^{in} \mu_s(i)\right]. \qquad (K.20)$$

The change in the Gibbs energy per solute molecule is

$$\Delta g = \frac{\Delta G}{N_s} = \mu_s - \sum_{i=1}^{m} x_i^{in} \mu_s(i)$$

$$= \left[\mu_s^* - \sum_{i=1}^{m} x_i^{in} \mu_s^*(i) \right] + k_B T \ln \rho_s \Lambda_s^3$$

$$- \sum_{i=1}^{m} x_i^{in} k_B T \ln \rho_i \Lambda_i^3, \qquad (K.21)$$

where we have used Eqs. (K.10) and (K.11). The three terms on the RHS of Eq. (K.21) correspond to the following three processes: (a) the intrinsic change in the Gibbs energy due to the change in the conformation [see Eq. (K.18)]; (b) the gain of the "liberation" Gibbs energy of a single solute s; this is the change in the Gibbs energy due to the release of a solute particle from a fixed position in the solvent w, and (c) the loss of the "liberation" Gibbs energy of all the solute species due to fixing their locations.

Using Eq. (K.12), we can rewrite Eq. (K.21) as

$$\Delta g = k_B T \sum x_1^{in} \ln x_i^{eq} - k_B T \sum_{i=1}^{m} x_i^{in} \ln x_i^{in}$$

$$= k_B T \ln \sum x_i^{in} \ln \frac{x_i^{eq}}{x_i^{in}} \le 0. \qquad (K.22)$$

The last inequality in Eq. (K.22) is valid for any two distributions. The equality holds if and only if $x_i^{in} = x_i^{eq}$ for all i. (This is also known as the Kullback–Leibler distance between the two distributions.) [See Papoulis and Pillain (2002); Ben-Naim (2008).]

Thus, we have shown that there exists a specific distribution x^{eq} [given by Eq. (K.12)] at which the Gibbs energy has a single minimum over all possible distributions x^{in}. We write this function as

$$G = F\left[T, P, N_w; x(\phi)\right], \tag{K.23}$$

where $x(\phi) = (x_1(\phi_1)), x_2(\phi_2), \ldots, x_n(\phi_n)$. This is the *composition vector* of the solute particles.

We have shown that the function $F(T, P, N_w; x(\phi))$ has a single minimum with respect to the discrete vector $x(\phi)$. However, we can go to the limit of a continuous distribution function, and interpret the vector $x(\phi)$ as a *function $x(\phi)$*. In that case, $F(T, P, N_w; x(\phi))$ is a *functional* of the continuous distribution $x(\phi)$.

Next, we show that $x(\phi)$ is related to the GEL.

In the same system characterized by (T, P, N_w, N_s) with $N_s \ll N_w$, we can ask for the probability of finding a single solute particle at any specific angle ϕ.

The probability density is given by

$$Pr(\phi) = \frac{\exp\left[-\beta f(\phi)\right]}{\int_0^{2\pi} \exp\left[-\beta f(\phi)\right] d\phi}, \tag{K.24}$$

where $f(\phi)$ is the Gibbs energy of a system at a given T, P, N_w, and one solute particle at a specific angle ϕ. This is the GEL of the solute having a single rotational degree of freedom. Thus, the GEL in this system is

$$G = f(T, P, N_w; \phi). \tag{K.25}$$

In a system characterized by T, P, N_w, N_s with $N_s \ll N_w$, the probability density $Pr(\phi)$ is the same as the function $x(\phi)$, where $x(\phi)d\phi$ is the mole fraction of solute molecules being at conformations between ϕ and $\phi + d\phi$.

Since $Pr(\phi)$ [or $x(\phi)$] is uniquely determined by the function $f(\phi)$, we can conclude that there exists an equilibrium GEL, denoted $f^{eq}(\phi)$, which minimizes the *functional* $G = \bar{F}(T, P, N_w, N_s; f(\phi))$, and the value of the Gibbs energy G at the minimum is the same as the one obtained for the equilibrium distribution $x^{eq}(\phi)$. [(We use $\bar{F}$ to distinguish this functional from F defined in Eq. (K.23)].

We have used the notation $f(\phi)$ and $F[x(\phi)]$ to distinguish between two different *functions*. Both are Gibbs energies. However, the first is a GEL, i.e. the Gibbs energy as a *function* of the angle ϕ. The second is a *functional*, i.e. the Gibbs energy as a functional of the function $x(\phi)$.

The second law of thermodynamics states that for a given environment (T, P, N_w, N_s), the *functional* F has a single (and hence global) minimum at an equilibrium distribution $x^{eq}(\phi)$ or, equivalently, at the equilibrium GEL, $f^{eq}(\phi)$. This means that, starting with any initial distribution of species, the system will relax to the equilibrium distribution. It does not make any statement about the number of minima or maxima in the GEL, or about the relative depths of the minima in the GEL.

The equilibrium distribution $x^{eq}(\phi)$ is related to the GEL $G = f(\phi)$ by Eq. (K.24). Therefore, we can also say that the Gibbs energy *functional* $G = \bar{F}[f(\phi)]$ has a unique minimum at the equilibrium GEL, corresponding to the equilibrium distribution $x^{eq}(\phi)$. It is now clear why we have used the different notation for the value of the Gibbs energy: the functional F and the GEL f. Using the same letter for the three, we would have the equation $G = G[T, P, N; G(\phi)]$

which could be potentially misleading since each of the three G's has a different meaning.

The generalization to protein is quite straightforward. Simply replace the single internal rotational angle ϕ by any number of rotational angles. Alternatively, reinterpret ϕ in the above discussion as the vector of all rotational angles ψ, as we have done in Chapter 6 and Appendix J.

Appendix L

Nonadditivity of the Gibbs Energy Function

In 1935, Kirkwood proposed the so-called superposition approximation [Kirkwood (1935)]. It states that the potential of mean force (PMF) for three or more particles may be approximated by the sum of pairwise PMFs. The PMF is defined as the change in the Helmholtz energy (in the T, V, N ensemble), or the Gibbs energy (in the T, P, N ensemble) associated with the process of bringing the three particles from infinite separation to the final configuration. [See Hill (1956); Münster (1974).] Thus, in the T, P, N ensemble,

$$W(R_1, R_2, R_3) = G(T, P, N; R_1, R_2, R_3)$$

$$-G(T, P, N; R_{ij} = \infty). \quad \text{(L.1)}$$

The question we pose here is: Under what conditions, if any, can one approximate $W(R_1, R_2, R_3)$ as a sum over pairs of PMFs?

$$W(R_1, R_2, R_3) = W(R_1, R_2) + W(R_1, R_3)$$
$$+ W(R_2, R_3). \qquad (L.2)$$

This is known as the Kirkwood superposition approximation (KSA). It was originally introduced by Kirkwood in the theory of liquids to achieve a closure to an integral equation for the pair correlation function. This approximation is intuitively very appealing. The reason is that such a pairwise-additive assumption is a good approximation for the *potential energy function* (Figure L.1) i.e.

$$U(R_1, R_2, R_3) \approx U(R_1, R_2) + U(R_1, R_3) + U(R_2, R_3).$$
$$(L.3)$$

In fact, the pairwise additivity of the potential function is exact for some systems, e.g. hard spheres (HSs), point charges,

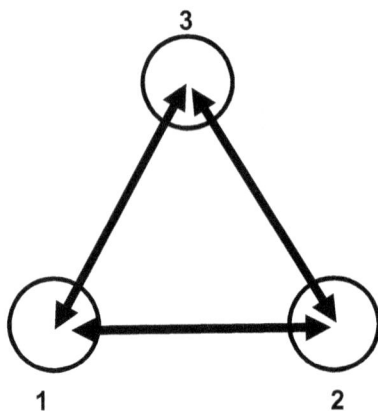

Fig. L.1. The energy potential change for bringing the three particles to a configuration (R_1, R_2, R_3), is the sum of the three energy changes for the process of bringing two particles to a distance as indicated by the double arrows.

point dipoles. It is also a good approximation for the total interaction energy among three or more simple, nonpolar molecules.

Perhaps Kirkwood was inspired by the additivity of the potential energy (L.3) to suggest the additivity of the PMF (L.2). Over the years, many authors attempted to improve upon this approximation [Münster (1974); Rice and Lekner (1965); Rice and Young (1967); Meeron (1957); Saltpeter (1958)]. Most of these attempts were aimed at improving the closure relation for the integral equation for the pair correlation function. Recently, an interesting new approach was suggested by Singer. Singer (2004) showed that the KSA may be obtained by applying the principle of maximum entropy [Rosenkratz (1989); Singer (2004)]. The question still remains: To what extent is the KSA a good approximation? In other words, is the KSA a good approximation, in itself, without any reference to a closure to an integral equation?

In the original article, Kirkwood used a probabilistic argument to justify the superposition approximation, namely that the probability of observing, say, three particles at a configuration R_1, R_2, R_3 is related to the product of the three pairwise probabilities. Equivalently, the superposition approximation may be formulated in terms of triplet and pair correlation functions:

$$g^{(3)}(R_1, R_2, R_3) \approx g^{(2)}(R_1, R_2)g^{(2)}(R_1, R_3)g^{(2)}(R_2, R_3).$$

$$(L.4)$$

The equivalency between (L.2) and (L.4) follows from the definition of the PMF and the corresponding correlation function for any number of particles:

$$W^{(n)}(R_1, \ldots, R_n) = -k_B T \ln g^{(n)}(R_1, \ldots, R_n). \quad \text{(L.5)}$$

Nowadays, it is not uncommon to encounter, especially in the biochemical literature, reference to the PMF as a *potential energy function*. As a "potential energy," one is inclined to apply the additivity assumption (L.3), without examining its validity. This nomenclature, though common, is unfortunate. The PMF is different from a *potential energy* in some fundamental properties. One is that the PMF is temperature-dependent, whereas the potential energy is approximately independent of temperature. The second is the nonadditivity of the PMF, even when there is an exact additivity of the potential energy.

In the rest of this appendix, we will present arguments showing the source of the nonadditivity of the PMF. We will show that the superposition approximation cannot be theoretically justified even when the potential energy of interactions is strictly additive. Therefore, we can conclude that the KSA is not an approximation at all. In Appendices M and N, we show a few examples to demonstrate the nonadditivity of the PMF.

L.1 The Source of the Nonadditivity of the PMF

The starting point of our argument is the following: for a classical system, we can write the triplet PMF as

$$\begin{aligned} W(R_1, R_2, R_3) &= U(R_1, R_2, R_3) + \delta G(R_1, R_2, R_3) \\ &= U(R_1, R_2, R_3) + \Delta G^*(R_1, R_2, R_3) \\ &\quad - 3\Delta G_s^*. \end{aligned} \quad \text{(L.6)}$$

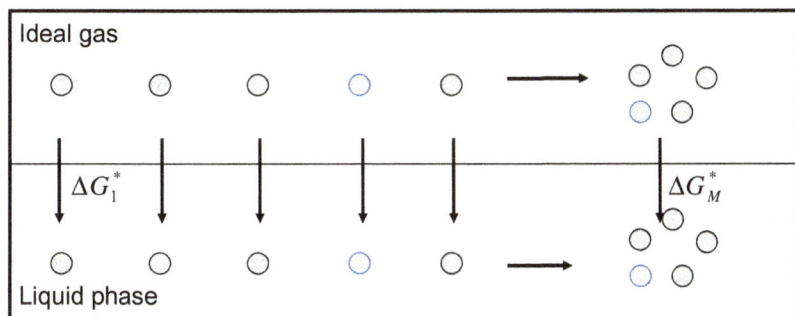

Fig. L.2. The relationship between Gibbs energy of the process of aggregation and the solvation Gibbs energies.

Equation (L.6) states that the work (here at T, P, N constants) of bringing three particles 1, 2 and 3 from infinite separations to the final configuration R_1, R_2, R_3 can be written as two terms: a *direct interaction energy*, $U(R_1, R_2, R_3)$, and a solvent-induced part, $\delta G(R_1, R_2, R_3)$. The latter can be rewritten as the difference in the solvation Gibbs energy of the triplet of particles at the configuration R_1, R_2, R_3, (Figure L.2), and three times the solvation Gibbs energy of one particle in the same solvent. Here, by "solvent," we mean all other particles in the system excluding the three particles 1, 2 and 3.

To highlight the source of the nonadditivity of the PMF, we will assume that the potential energy in (L.6) is pairwise-additive (approximately or exactly). We will show below that the nonadditivity of the PMF originates from the *solvation* Gibbs energy $\Delta G^*(R_1, R_2, R_3)$.

Note that when the solvent density tends to zero, all the solvation Gibbs energies on the RHS of Eq. (L.6) will tend to zero. At this limit, the PMF becomes identical with the potential energy. It is perhaps this limiting example that inspired Kirkwood as well as many others to adopt the supposition approximation for the PMF. Unfortunately, whenever a solvent is present, even a solvent consisting of a single molecule, the additivity assumption of the PMF becomes invalid.

We first show the source of the nonadditive for the simplest solvent: a single water molecule. We also use the T, V, N ensemble for convenience. The conclusions are also valid for the T, P, N ensemble.

The solvation Helmholtz energy ΔA^* of M particles at any specific configuration R^M is defined as

$$\Delta A^*(R^M) = -k_B T \ln_0 \left\langle \exp\left[-\beta B\left(R^M\right)\right]\right\rangle, \qquad (L.7)$$

where k_B is the Boltzmann constant, T is the absolute temperature, $\beta = (k_B T)^{-1}$ and $B(R^M)$ is the *binding energy*, defined by

$$B(R^M) = \sum_{k=1}^{M} U(R_k, X_w). \qquad (L.8)$$

The average in Eq. (L.7) is over all configurations of the water molecule in the T, V, N ensemble with the probability distribution

$$P(X_w) = \frac{1}{V(8\pi^2)}. \qquad (L.9)$$

Assuming that the total interaction energy in the system is pairwise-additive, we can write the Helmholtz energy function as

$$A(\boldsymbol{R}^M) = \sum_{i<j} U(\boldsymbol{R}_i, \boldsymbol{R}_j) - k_B T \ln \int \prod_{k=1}^{M}$$
$$\times \exp[-\beta U(\boldsymbol{R}_k, \boldsymbol{X}_w)] P(\boldsymbol{X}_w) d\boldsymbol{X}_w - M \Delta A_s^*.$$
$$(\text{L}.10)$$

The integration in Eq. (L.10) is over all possible configurations of the water molecule. We can now see the main difference between the potential energy function $U(\boldsymbol{R}^M)$ and the Helmholtz energy function $A(\boldsymbol{R}^M)$.

The total interaction energy in a system of M particles (here $M = 3$) is the sum of the interaction energies of all pairs of atoms. These are indicated by bold double arrows in Figure L.3. The interactions between the M particles and the water molecule are indicated by the dashed double arrows [Figure L.3(a)], and for the more general case in [Figure L.3(b)].

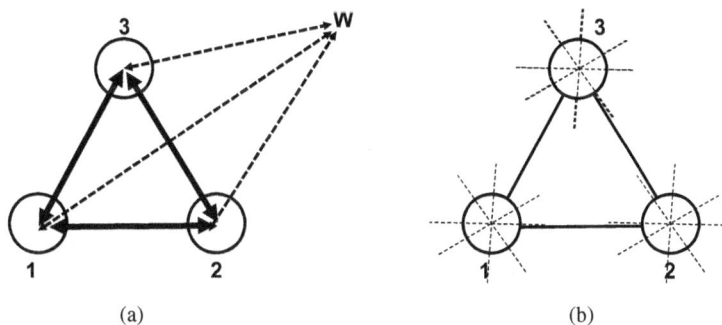

(a)　　　　　　　　(b)

Fig. L.3. The total interaction energy in a system of M particles (here $M = 3$), is the sum of the interaction energies between all pairs of atoms. There are indicated by bold double arrows. The interactions between the M particles and the water molecule are indicated by the dashed double arrows in (a), and for the more general case in (b).

The assumption of pairwise additivity of the potential function (whether exact or approximate) is indicated in [Figure L.3(a)] as bold double arrows connecting all the M particles. On the other hand, the second term on the RHS of Eq. (L.10) contains only "lines of interactions" between the M particles and the water molecule. These are shown as dashed double arrows in [Figure L.3(a)].

Note that the second term on the RHS of Eq. (L.10) has no component which is pairwise-additive in the sense that $U(R^M)$ has. Therefore, one cannot assume that the Helmholtz energy function is pairwise-additive, unless the second term on the RHS of Eq. (L.10) is negligible, which means that the Helmholtz energy function reduces to the potential energy function.

The conclusion reached above is valid for any solvent — not necessarily the simplest "solvent" discussed above. In the more general case, the Helmholtz energy function is written as

$$A\left(R^M\right) = U\left(R^M\right) + \Delta A^*\left(R^M\right) - M\,\Delta A_s^*$$

$$= U\left(R^M\right) - k_B T \ln$$

$$\times \int \cdots \int \exp\left[-\beta B\left(R^M\right)\right]$$

$$\times P\left(X^N\right) dX^N - M\,\Delta A_s^*, \qquad (L.11)$$

where $P(X^N)$ is the configurational distribution of the solvent molecules.

Again, we see that the function $U(R^M)$ has lines of interactions between all pairs of particles. On the other hand, the solvation Gibbs energy has only lines of inter-actions connecting the solute particles to water molecules [Figure L.3(b)]. The averaging over all the configurations of the solvent molecules has no effect on this conclusion.

Thus, the general argument is that although the processes of bringing the three particles in vacuum and in the liquid are the same processes, the works associated with these two processes are very different. The very rewriting of the PMF in the form (L.6), or the more general form (L.11), reveals the inadequacy of the KSA. In the limit of low solvent density, $\Delta G^*(R^M)$ may be shown to be factorizable into a product of solvation Gibbs energies of the M *single* particles. In this limit, the last two terms on the RHS of Eq. (2.6) cancel out and we are left with the potential energy, which to a good approximation may be assumed to be pairwise-additive.

Next, we present an exact example demonstrating the nonadditivity of the PMF. In Appendix M, we give another example of nonadditivity of the Gibbs energy.

Three Hard Spheres at R_1, R_2, R_3, and a Solvent Consisting of a Single Hard Sphere

This is the simplest case of solvation of the triplet of hard sphere (HS) particles in a one-hard-sphere "solvent," denoted w. The solvation Helmholtz energy in this case is

$$
\begin{aligned}
&\exp\left[-\beta\Delta A^*\left(R_1, R_2, R_3\right)\right] \\
&= \left\langle \exp\left[-\beta B(R_1, R_2, R_3)\right]\right\rangle \\
&= \int \frac{dR_W}{V} \exp\left[-\beta B(R_1) - \beta B(R_2) - \beta B(R_3)\right]
\end{aligned}
$$

$$= \frac{1}{V} \int \frac{d\boldsymbol{R}_W}{V} \exp\left[-\beta U(R_{1W})\right.$$

$$\left. - \beta U(R_{2W}) - \beta B(R_{3W})\right]$$

$$= \frac{V - V^{EX}(\boldsymbol{R}_1, \boldsymbol{R}_2, \boldsymbol{R}_3)}{V} \tag{L.12}$$

In the second step on the RHS of Eq. (L.12), we have written the average quantity explicitly. Here, the probability density of the solvent molecule is simply

$$\Pr(\boldsymbol{R}_W) = \frac{1}{V}. \tag{L.13}$$

Since there is only one solvent molecule, denoted w, the binding energy reduces to the solute–solvent interaction energy. We next use the property of the HS interaction potential, which makes the integrand zero whenever the solvent molecule penetrates into the excluded volume (V^{EX}) of the triplet of particles, and unity otherwise. The resulting expression is the RHS of Eq. (L.12).

Figure L.4 shows three configurations of the three particles and the corresponding excluded volumes. For simplicity, we assume that the configuration R_1, R_2, R_3 is an equilateral triangle. Since V is much larger than V^{EX}, we can expand ΔA^*, to first order in V^{EX}/V, to obtain

$$\Delta A^*(\boldsymbol{R}_1, \boldsymbol{R}_2, \boldsymbol{R}_3) = -k_B T \ln\left(1 - \frac{V^{EX}(\boldsymbol{R}_1, \boldsymbol{R}_2, \boldsymbol{R}_3)}{V}\right)$$

$$\approx \frac{k_B T}{V} V^{EX}(\boldsymbol{R}_1, \boldsymbol{R}_2, \boldsymbol{R}_3), \tag{L.14}$$

and similarly for a pair of particles at R_1, R_2:

$$\Delta A^*(R_1, R_2) \approx \frac{k_B T}{V} V^{EX}(R_1, R_2). \qquad (L.15)$$

The assumption of pairwise additivity is equivalent to the equality of

$$V^{EX}(R_1, R_2, R_3) = V^{EX}(R_1, R_2)$$
$$+ V^{EX}(R_1, R_3) + V^{EX}(R_2, R_3). \qquad (L.16)$$

Clearly, an equality of this kind does not exist, as can be seen from Figure L.4. Note that when the particles are sufficiently far apart, such that the excluded volume around the particles is the sum of the excluded volume of each particle,

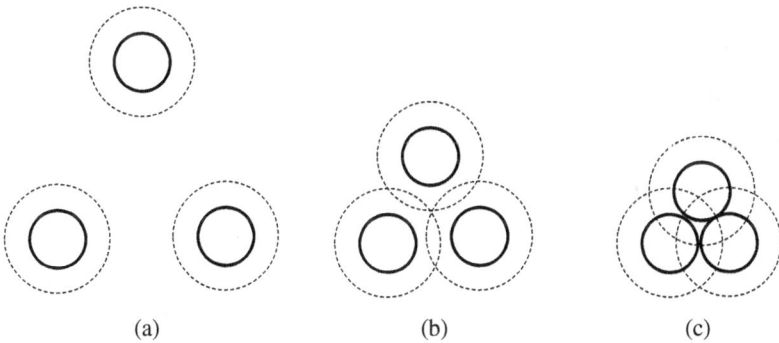

(a)   (b)   (c)

Fig. L.4. Three configurations of three spherical particles (here depicted as three disks). At large distances, there is no overlap in the excluded volume about each particle (a). At closed distances, there is some overlap (b), and the maximum overlap is shown in the configuration (c).

we have

$$V^{\text{EX}}(R_1, R_2, R_3) = 3V^{\text{EX}}(R_1), \qquad \text{(L.17)}$$

$$V^{\text{EX}}(R_1, R_2) = 2V^{\text{EX}}(R_1). \qquad \text{(L.18)}$$

In this case, the equality (L.16) certainly does not hold. However, the solvent-induced contribution δ_G and the interaction energy are zero, and this case is of no interest.

L.2 Conclusion

We have shown that the KSA cannot be considered either a good or a bad approximation, for the PMF between any number of particles and in any solvent. The reason for that conclusion is that the extent of nonadditivity of the solvation Gibbs or Helmholtz energy is, in general, of the same order of magnitude as the Gibbs or Helmholtz energy of solvation itself.

Unfortunately, the KSA is used, either explicitly or implicitly, in many studies of aqueous solutions, particularly the study of protein folding and protein–protein association. Here, it is more common to refer to the potential energy function as the EL, and to the potential of mean force as the GEL.

In processes such as protein folding or protein–protein association, it is believed that the solvation Gibbs energy term dominates the GEL. Therefore, we can conclude that even when the EL is strictly pairwise-additive, the GEL can never be assumed, not even approximated, as a pairwise-additive function. This conclusion is valid for any solvent, or a mixture of solvents of any composition.

Appendix M

Nonadditivity in Binding Systems

In this appendix, we present an exactly solvable model to demonstrate the source as well as the magnitude of the nonadditivity of the Gibbs energy.

M.1 The Model and Its Partition Function

Consider a molecule having three binding sites. It can be in two different conformations, denoted L and H (Figure M.1). The sites are identical, and each site can be either empty or occupied by a ligand.

The grand partition function (GPF) for a single molecule is

$$
\begin{aligned}
\xi &= Q(0) + Q(1)\lambda + Q(2)\lambda^2 + Q(3)\lambda^3 \\
&= Q(0,0,0) + [Q(a,0,0) + Q(0,b,0) + Q(0,0,c)]\lambda \\
&\quad + [Q(a,b,0) + Q(0,b,c) + Q(a,0,c)]\lambda^2 \\
&\quad + Q(a,b,c)\lambda^3 \\
&= Q(0,0,0) + 3Q(1) + 3Q(2)\lambda^2 + Q(3)\lambda^3. \quad \text{(M.1)}
\end{aligned}
$$

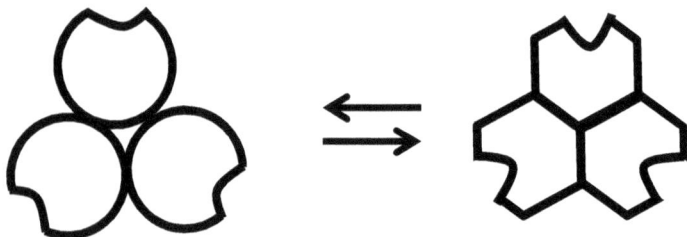

Fig. M.1. Tow conformations of a three-binding-sites molecule.

In the first equality, the GPF is written in the most general form (for the case $m = 3$). Here, $Q(l)$ is the canonical partition function (PF) for a system with l ligands. The second form is used whenever the sites are all *different*, here denoted a, b and c. Clearly, in this case, there are three PFs for singly occupied systems, denoted $(a, 0, 0)$, $(0, b, 0)$ and $(0, 0, c)$, and three different PFs for doubly occupied systems. The third form is used when the three sites are *identical*. Thus, instead of writing the *specific* PFs, $Q(1, 0, 0)Q(0, 1, 0)$ and $Q(0, 0, 1)$, which have the same value, denoted $Q(1)$, we simply take three times this quantity. Similarly, for the doubly occupied system, we write $Q(2)$, and for the triply occupied system, we write $Q(3)$. λ is the absolute activity of the ligand.

We next define the energies E_L and E_H for the two states of the molecule L and H, respectively. We also denote by U_L and U_H the binding energies of the ligand to a site when the molecule is in states L and H, respectively. Denote by $U(1, 1)$, the interaction energy between the two ligands occupying two sites. We assume that this interaction energy is independent of which two sites are occupied and independent of the state (L or H) of the molecule. We also assume that the interaction energy between three ligands occupying three sites is strictly pairwise-additive, i.e.

$$U(a, b, c) = U(a, b) + U(a, c) + U(b, c). \quad \text{(M.2)}$$

Since the sites are identical, we can rewrite Eq. (M.2) as

$$U(1, 1, 1) = 3U(1, 1). \quad \text{(M.3)}$$

We use the notation 1 or 0 for occupied or empty sites, respectively.

We denote

$$Q_L = \exp(-\beta E_L), \quad Q_H = \exp(-\beta E_H), \quad \text{(M.4)}$$

$$q_L = \exp(-\beta U_L), \quad q_H = \exp(-\beta U_H), \quad \text{(M.5)}$$

$$S = \exp\left[-\beta U(1, 1)\right]. \quad \text{(M.6)}$$

With these notations, we can rewrite the GPF in Eq. (M.1) as

$$\begin{aligned}
\xi &= Q(0) + 3Q(1)\lambda + 3Q(2)\lambda^2 + Q(3)\lambda^3 \\
&= \left[Q_L + Q_H\right] + 3\left[Q_L q_L + Q_H q_H\right]\lambda \\
&\quad + 3\left[Q_L q_L^2 + Q_H q_H^2\right] S\lambda^2 + \left[Q_L q_L^3 + Q_H q_H^3\right] S^3 \lambda^3.
\end{aligned}$$
$$\text{(M.7)}$$

This is an explicit expression for the GPF of a single adsorbent molecule in terms of the energies defined for this system: $E_L, E_H, U_L, U_H, U(1, 1)$. With this GPF, one can calculate the adsorption isotherm for such a system [see Ben-Naim (2001)]. We will be interested here in some thermodynamic aspects of this system which we will evaluate below.

M.2 Probabilities and Correlations

All the probabilities that we need are obtainable from the PF in Eq. (M.7). For our purpose, we need the following probabilities:

The probability of finding a specific site occupied:

$$P_1 = P(1, \_, \_) = \frac{(Q_L q_L + Q_H q_H) \lambda}{\xi}. \qquad \text{(M.8)}$$

Note that $P(1, \_, \_)$ means the probability that the first site is occupied, and the states of the other two sites are unspecified.

Because of the equivalence of the three sites, we have the equality

$$P_1 = P(1, \_, \_) = P(\_, 1, \_) = P(\_, \_, 1). \qquad \text{(M.9)}$$

In each of these probabilities, one specific site is occupied. The probability of finding a molecule with *one* site occupied is simply $3P_1$.

The probability of finding two specific sites occupied is given by

$$P_{12} = P(1, 1, \_) = \frac{(Q_L q_L^2 + Q_H q_H^2) S \lambda^2}{\xi}, \qquad \text{(M.10)}$$

and since the sites are equivalent we have

$$P_{12} = P(1, 1, \_) = P(1, \_, 1) = P(\_, 1, 1). \qquad \text{(M.11)}$$

Finally, the probability of finding the three sites occupied is

$$P_{123} = P(1, 1, 1) = \frac{(Q_L q_L^3 + Q_H q_H^3) S^3 \lambda^3}{\xi}. \qquad \text{(M.12)}$$

Note that all the probabilities above depend on the absolute activity of the ligand.

The two correlations of interest are

$$g^2(1, 1) = \frac{P_{12}}{P_1^2}, \qquad \text{(M.13)}$$

$$g^3(1, 1, 1) = \frac{P_{123}}{P_1^3}. \qquad \text{(M.14)}$$

In the limit of very low concentration of the ligand, $\lambda \to 0$, we have for the two correlations

$$g^2(1, 1) = \frac{(Q_L + Q_H)\left(Q_L q_L^2 + Q_H q_H^2\right)}{\left(Q_L q_L + Q_H q_H\right)^2} S = y^{(2)} S,$$

$$\text{(M.15)}$$

$$g^3(1, 1, 1) = \frac{(Q_L + Q_H)^2 \left(Q_L q_L^3 + Q_H q_H^3\right)}{\left(Q_L q_L + Q_H q_H\right)^3} S^3 = y^{(3)} S^3.$$

$$\text{(M.16)}$$

We further define the two parameters

$$K = \frac{Q_H}{Q_L}, \quad h = \frac{q_H}{q_L}. \qquad \text{(M.17)}$$

With these parameters, we can rewrite the two correlations as

$$g^2(1, 1) = \frac{\left(1 + Kh^2\right)(1 + K)}{\left(1 + Kh\right)^2} S$$

$$= \left[1 + \frac{K\left(h - 1\right)^2}{\left(1 + Kh\right)^2}\right] S \qquad \text{(M.18)}$$

$$g^3(1, 1, 1) = \frac{\left(1 + Kh^3\right)\left(1 + K\right)^2}{\left(1 + Kh\right)^3} S^3$$

$$= \left[1 + \frac{K\left(h - 1\right)^2 \left(2Kh + K + h + 2\right)}{\left(1 + Kh\right)^3}\right] S^3.$$

$$(\text{M}.19)$$

In Eqs. (M.15) and (M.16), we also defined the *direct* correlations S and S^3, which are due to the direct interactions between the ligands, and the *indirect* correlations, denoted $y^{(2)}$ and $y^{(3)}$, which are due to indirect interactions between the ligands.

The latter occur whenever the binding of ligands changes the distribution of the molecules in the L and H forms. [For more details, see Ben-Naim (2001).] It is easy to see from Eqs. (M.18) and (M.19) that "either $K = 0$ or $h = 1$" is a necessary and sufficient condition for no indirect correlations, i.e. $y^{(2)} = y^{(3)} = 1$. $K = 0$ or $K = \infty$ means that one conformation is infinitely more stable than the second:

$$K = \frac{Q_H}{Q_L} = \exp\left[-\beta\left(E_H - E_L\right)\right]. \qquad (\text{M}.20)$$

For $E_H - E_L \to \infty$, $K \to 0$, and for $E_H - E_L \to -\infty$, $K \to \infty$. In both of these extreme cases, the ligand cannot produce any conformational change in the adsorbent molecule. For any other value of $K \neq 0$ and $K \neq \infty$, $h = 1$ means that U_L equals U_H. In this case, the ligand cannot affect the conformational equilibrium of the adsorbent molecule. When $h > 1$, the ligand stabilizes the H form, and for $h < 1$, it stabilizes the L form.

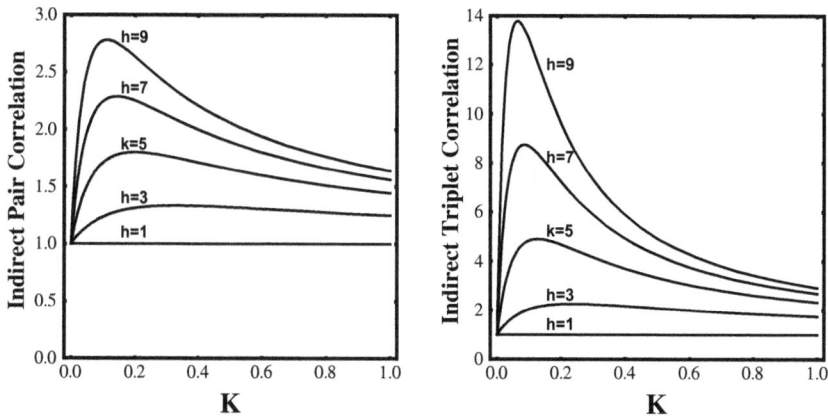

Fig. M.2. The indirect pair and triplet correlations as a function K for different values of h.

Figure M.2 shows the indirect correlation functions $y^{(2)}$ and $y^{(3)}$ as a function of K for different values of h. We see that both the pair and the triplet correlations are always *positive*, i.e.

$$y^{(2)} > 1 \quad \text{and} \quad y^{(3)} > 1.$$

We next examine the pair and triplet correlation functions defined in Eqs. (M.15) and (M.16). Because of the symmetry of the adsorbed molecule, the interaction energy between the ligands on each pair of sites is independent of the state (L or H) of the molecule. We choose the direct correlation, such that

$$S = \exp(-\beta J), \tag{M.21}$$

where $J > 0$ corresponds to the repulsion interaction, and hence the direct correlation $S < 1$. We choose the case $h = 9$ from Figure M.2 and examine the effect of changing $J = 0, 1, 2, 3, 4$. The larger the value of J, the larger the

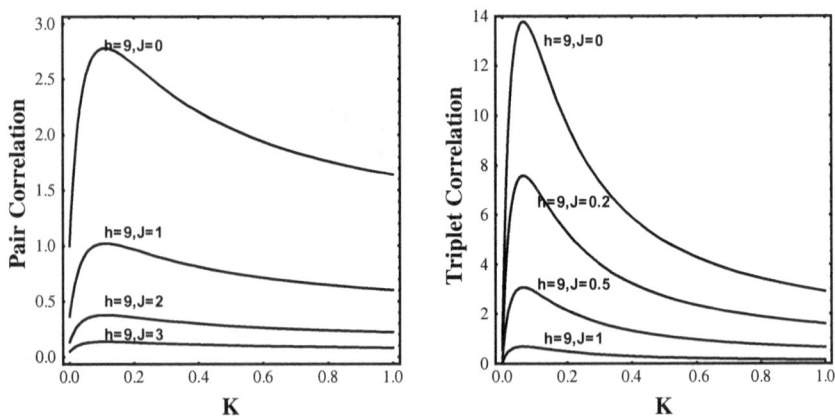

Fig. M.3. The indirect pair and triplet correlations as a function of K for different values of J and $h = 9$.

negative pair correlation. Figure M.3 shows the pair and triplet correlations for $h = 9$ and various values of J. We see that for $J = 0$, the correlations are larger than 1, and as we increase J, the correlations become more and more *negative*, i.e.

$$g^{(2)} < 1, \quad g^{(3)} < 1.$$

We next turn to the PMF and the extent of the nonadditivity of the PMF. We have chosen the direct interactions to be strictly additive, in the sense of Eq. (M.3). Therefore, we can focus only on the indirect part of the PMF.

The two relevant quantities are

$$W^{(2)}(1, 1) = -k_B T \log g^{(2)}(1, 1), \qquad (\text{M.22})$$
$$W^{(3)}(1, 1, 1) = -k_B T \log g^{(3)}(1, 1, 1). \qquad (\text{M.23})$$

Figure M.4 shows the indirect part of the PMF defined in Eqs. (M.22) and (M.23) (with $S = 1$). These quantities

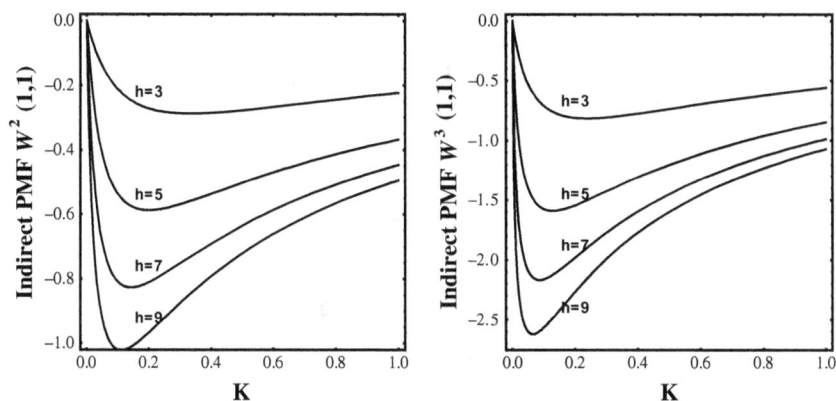

Fig. M.4. The indirect part of the PMF as a function of K for different values of h.

correspond to the indirect work for the reactions:

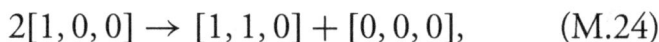

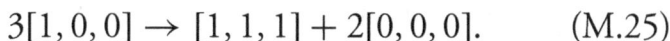

$$2[1, 0, 0] \rightarrow [1, 1, 0] + [0, 0, 0], \qquad (M.24)$$

$$3[1, 0, 0] \rightarrow [1, 1, 1] + 2[0, 0, 0]. \qquad (M.25)$$

The first corresponds to the process of converting two singly occupied molecules into one empty and one doubly occupied molecule. The second corresponds to the process of converting three singly occupied molecules into one triply occupied molecule and two empty molecules.

Note that all the values of $W^{(2)}$ and $W^{(3)}$ in Figure M.4 are negative.

Finally, in Figure M.5, we show the nonadditivity of the PMF defined by

$$\text{NONADD} = W^{(3)}(1, 1, 1) - 3W^{(2)}(1, 1). \qquad (M.26)$$

Note that since the direct interaction energy between the ligands is additive [Eq. (M.3)], the nonadditivity

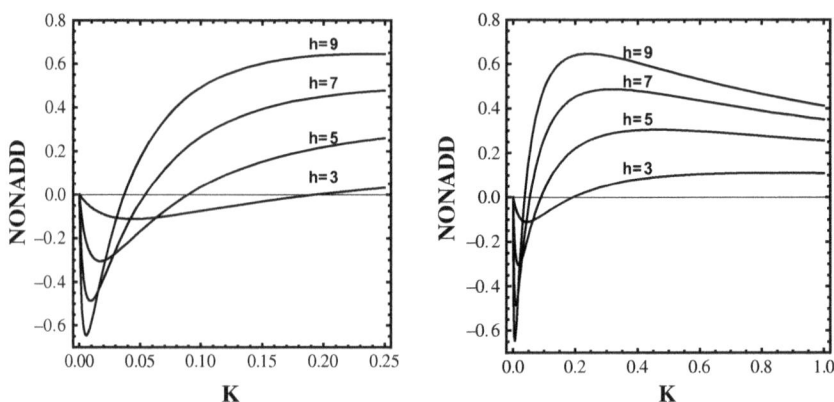

Fig. M.5. The extent of nonadditivity as a function of K for different values of h.

(NONADD) defined in Eq. (M.26) is essentially the nonadditivity of the *indirect* PMF. Figure M.5(a) shows the values of NONADD for $h = 3, 5, 7, 9$ in the range of $0 \leq K \leq 0.2$, and Figure M.5(b) shows the values of NONADD for an extended range of K.

The most important finding here is that the nonadditivity of the potential of average force can be either positive or negative, and its order of magnitude is not much different from the values of $W^{(2)}$ and $W^{(3)}$.

Appendix N

Frustration in Three Dipoles on a Lattice: An Informational-Theoretical Approach

In this appendix, we discuss the case of three dipoles (or any three random variables), each attaining one of two values: up and down or zero and one. We first calculate the pair and triplet PMFs for this system. We next reinterpret the same quantities using the concept of mutual information (MI).

N.1 The Three Dipoles at the Vertices of Regular Triangles

Consider the system of three dipoles denoted X_1, X_2, X_3 shown in Figure 7.10. The dipoles are located at the vertices of a regular triangle in the xy plane. Each dipole can be oriented either "up" or "down" along the z-axis.

We denote by x_i the *state* of the dipole X_i, and we assign the values $x_i = 1$ and $x_i = -1$ for the states "up" and "down," respectively.

The energy of the system being in state (x_1, x_2, x_3) is assumed to be pairwise-additive, i.e.

$$E(x_1, x_2, x_3) = -J(x_1x_2 + x_1x_3 + x_2x_3). \qquad \text{(N.1)}$$

All the probabilities for this system can be derived from the relationship:

$$P(x_1, x_2, x_3) = \frac{\exp\left[-\beta E\,(x_1, x_2, x_3)\right]}{Z_3}, \qquad \text{(N.2)}$$

where Z_3 is defined by

$$Z_3 = \sum_{x_1=\pm1} \sum_{x_2=\pm1} \sum_{x_3=\pm1} \exp\left[-\beta E\,(x_1, x_2, x_3)\right]. \quad \text{(N.3)}$$

The sum in Eq. (N.3) is overall the x_i, attaining the values $+1$ and $-1 \cdot \beta = (k_B T)^{-1}$, where k_B is the Boltzmann constant and T the absolute temperature. Altogether there are eight states, in two of which the energy is $E = -3\beta J$ [all the dipoles in the same direction; Figure 7.10(b)], and there are six configurations [one up and two down, or one down and two up; Figure 7.10(c)] with energy $E = \beta J$. Thus, for this system, the normalization constant in Eq. (N.2) is

$$Z_3 = 2 \exp{(3\beta J)} + 6 \exp{(-\beta J)}. \qquad \text{(N.4)}$$

Clearly, for electric dipoles ($J < 1$), the probability of a configuration of the type [Figure 7.10(b)] is lower that the probability of a configuration of the type [Figure 7.10(c)].

We now define the triplet MI by

$$I_3(X_1, X_2, X_3) = \sum \sum \sum P(x_1, x_2, x_3)$$

$$\times \log \frac{P(x_1, x_2, x_3)}{P(x_1)P(x_2)P(x_3)}. \qquad (N.5)$$

This is a straightforward generalization of the MI defined for two random variables by

$$I_2(X_1, X_2) = \sum \sum P(x_1, x_2) \log \frac{P(x_1, x_2)}{P(x_1)P(x_2)}$$

$$= \sum \sum P(x_1, x_2) \log g(x_1, x_2), \qquad (N.6)$$

where $g(x_1, x_2)$ is the correlation between the two events $X_1 = x_1$ and $X_2 = x_2$. Here, we use the natural logarithm to relate the MI to the analog of the PMF. In information entropy, one normally uses the base 2 for the logarithm. Note also that the generalization of MI in Eq. (N.4) is different from the one used in information theory. For instance, Matsuda (2000) defines the triplet MI by

$$F_3(X_1, X_2, X_3) = -I(X_1, X_2 | X_3) + I(X_1, X_2). \qquad (N.7)$$

This quantity measures the change in the MI between X_1 and X_2 when we "know" X_3. In the definition (N.7), the quantity $F_3(X_1, X_2, X_3)$ is not symmetric with respect to the three variables X_1, X_2, X_3 in Matsuda, our $F_3(X_1, X_2, X_3)$ is denoted $I_3(X_1, X_2, X_3)$. We now show that F_3 is symmetric with respect to any permutation of the variables X_1, X_2, X_3, and also show the relationship between F_3 and I_3.

The quantity $I(X_1, X_2 | X_3)$ is defined by

$$I(X_1, X_2 | X_3) = \sum_{x_3} P(x_3)\, I(X_1, X_2 | X_3 = x_3). \quad \text{(N.8)}$$

Here, $I(X_1, X_2 | X_3 = x_3)$ is the MI between the X_1 and X_2 conditions given a *specific* value of $X_3 = x_3$. To obtain $I(X_1, X_2 | X_3)$, we simply take the average over all possible values of x_3.

We now rewrite F_3 defined in Eq. (N.7) by using the probabilities of the various events. Thus, from Eqs. (N.7) and (N.8), we have

$$\begin{aligned}
F_3(X_1, X_2, X_3) = {}&- \sum_{x_3} P(x_3) \sum_{x_1, x_2} P(x_1, x_2 | x_3) \\
&\times \log \frac{P(x_1, x_2 | x_3)}{P(x_1 | x_3)\, P(x_2 | x_3)} \\
&+ \sum_{x_1, x_2} P(x_1, x_2) \log \frac{P(x_1, x_2)}{P(x_1)\, P(x_2)}.
\end{aligned}$$

$$\text{(N.9)}$$

$P(x_1)$ and $P(x_i, x_j)$ are the marginal distributions, i.e.

$$P(x_1, x_2) = \sum_{x_3} P(x_1, x_2, x_3), \quad \text{(N.10)}$$

$$P(x_1) = \sum_{x_2} P(x_1, x_2). \quad \text{(N.11)}$$

We can rewrite Eq. (N.9) as

$$F_3(X_1, X_2, X_3) = -\sum_{x_1, x_2, x_3} P(x_1, x_2, x_3)$$

$$\times \log \frac{P(x_1, x_2, x_3)}{P(x_3)} \frac{P(x_3)}{P(x_1, x_3)} \frac{P(x_3)}{P(x_2, x_3)}$$

$$+ \sum_{x_1, x_2, x_3} P(x_1, x_2, x_3) \log \frac{P(x_1, x_2)}{P(x_1) P(x_2)}$$

$$= -\sum_{x_1, x_2, x_3} P(x_1, x_2, x_3)$$

$$\times \log \frac{P(x_1, x_2, x_3) P(x_1) P(x_2) P(x_3)}{P(x_1, x_2) P(x_1, x_3) P(x_2, x_3)}.$$

$$(N.12)$$

We see from the last form of Eq. (N.12) that the quantity $F_3(X_1, X_2, X_3)$ is symmetrical with respect to the variables $X_1 X_2 X_3$. Therefore, we can extend the definition (N.7) as

$$F_3(X_1, X_2, X_3) = -I(X_1, X_2 | X_3) + I(X_1, X_2)$$

$$= -I(X_1, X_3 | X_2) + I(X_1, X_3)$$

$$= -I(X_2, X_3 | X_1) + I(X_2, X_3). \quad (N.13)$$

The equalities in (N.13) mean that given any of the variables $X_1 X_2$ and X_3, its effect on the MI between the other two variables is independent of the given variable.

We next relate $F_3(X_1, X_2, X_3)$ to the pair and triplet correlations, and to the corresponding PMF.

Using the definitions of the pair and triplet correlations,

$$g(x_1, x_2) = \frac{P(x_1, x_2)}{P(x_1) P(x_2)}, \quad (\text{N.14})$$

and similarly for (x_1, x_3) and (x_2, x_3)

$$g(x_1, x_2, x_3) = \frac{P(x_1, x_2, x_3)}{P(x_1) P(x_2) P(x_3)}. \quad (\text{N.15})$$

We can rewrite F_3 in Eq. (N.12) as

$$F_3(X_1, X_2, X_3) = - \sum_{x_1, x_2, x_3} P(x_1, x_2, x_3) \log g(x_1, x_2, x_3)$$

$$+ \sum_{x_1, x_2} P(x_1, x_2) \log g(x_1, x_2)$$

$$+ \sum_{x_1, x_3} P(x_1, x_3) \log g(x_1, x_3)$$

$$+ \sum_{x_2, x_3} P(x_2, x_3) \log g(x_2, x_3)$$

$$= -I_3(X_1, X_2, X_3) + I_2(X_1, X_2)$$

$$+ I_2(X_1, X_3) + I_2(X_2, X_3). \quad (\text{N.16})$$

Thus, we have related $F_3(X_1, X_2, X_3)$ to the pair and triplet MI. Note that all the pairs of MI are defined in the system of three (not two) dipoles.

We can now relate $F_3(X_1, X_2, X_3)$ to the extent of deviations from the KSA. This can be done either in terms of the correlations functions or in terms of the PMF, i.e.

$$F_3(X_1, X_2, X_3) = - \sum_{x_1, x_2, x_3} P(x_1, x_2, x_3)$$

$$\times \log \frac{g(x_1, x_2, x_3)}{g(x_1, x_2)\, g(x_1, x_3)\, g(x_2, x_3)}.$$

$$(N.17)$$

The KSA may be defined as $\frac{g(x_1, x_2, x_3)}{g(x_1, x_2)g(x_1, x_3)g(x_2, x_3)} = 1$, i.e. the triplet correlation is the product of the three pair correlations; see below for an alternative but equivalent definition.

If the KSA holds for *each configuration* x_1, x_2, x_3, then Eq. (N.17) reduces to

$$F_3(X_1, X_2, X_3) = 0. \qquad (N.18)$$

Equivalently, we use the relationship between the correlation g and the PMF,

$$W = -k_B T \log g, \qquad (N.19)$$

which applies to any number of dipoles. With Eq. (N.19), we can rewrite Eq. (N.17) as

$$F_3(X_1, X_2, X_3) = \beta \sum_{x_1, x_2, x_3} P(x_1, x_2, x_3)\, [W(x_1, x_2, x_3)$$

$$- W(x_1, x_2) - W(x_1, x_3) - W(x_2, x_3)].$$

$$(N.20)$$

We define the deviation from the KSA by

$$\delta W(x_1, x_2, x_3) = W(x_1, x_2, x_3) - W(x_1, x_2)$$

$$- W(x_1, x_3) - W(x_2, x_3). \qquad (N.21)$$

Hence, we have

$$F_3(X_1, X_2, X_3) = \beta \langle \delta W \rangle. \qquad \text{(N.22)}$$

Thus, the quantity F_3 is an *average* of δW over all possible configurations of the three dipoles. Again, we note that if the KSA holds for any configuration $x_1 x_2 x_3$, then $F_3(X_1, X_2, X_3) = 0$.

N.2 Some Numerical Results

Before we discuss the case of three dipoles, it is instructive to examine the simpler case of two dipoles, as shown in Figure 7.9. In this case, there are four possible configurations (up–up, down–down, up–down and down–up). The probabilities are obtained from the pair distribution

$$P^{(2)}(x_1, x_2) = \frac{\exp\left[-\beta J(x_1, x_2)\right]}{Z_2}, \qquad \text{(N.23)}$$

where

$$Z_2 = \sum_{x_1 = \pm 1} \sum_{x_2 = \pm 1} \exp\left[-\beta J(x_1, x_2)\right]. \qquad \text{(N.24)}$$

The singlet probability is

$$P^{(2)}(x_1) = \sum_{x_2} P^{(2)}(x_1, x_2). \qquad \text{(N.25)}$$

In this case, $P^{(2)}(1) = P^{(2)}(-1) = 0.5$, as expected. We note also that singlet probabilities are independent of J. The superscript (2) is added to emphasize that we have only two dipoles.

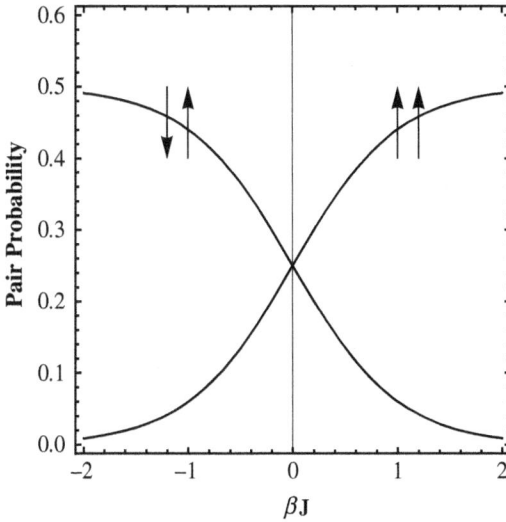

Fig. N.1. The pair probability as a function of βJ for the two configurations of the dipoles.

Figure N.1 shows the pair distributions defined in Eqs. (N.24). For $\beta J \to \infty$, the up–up (or down–up) has probability 0.

Figure N.2 shows the pair correlation for each specific configuration. We see that for $J = 1$, the pair correlation for the up–up (and down–down) is *positive*, i.e. $g(1, 1) > 1$ and $g(-1, -1) > 1$. On the other hand, for the up–down (and down–up), the correlation is negative, i.e. $g(1, 1) < 1$ and $g(-1, -1) < 1$. The pair correlation is defined by

$$g(x_1, x_2) = \frac{P(x_1, x_2)}{P(x_1)\, P(x_2)}. \tag{N.26}$$

Note that $g > 1$ and $g < 1$ correspond to the PMF being positive and negative, respectively. These are shown in [Figure N.2(b)].

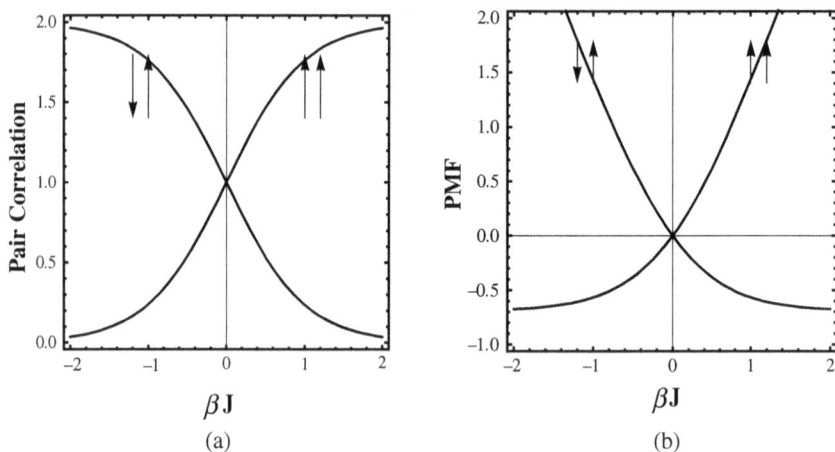

Fig. N.2. The pair correlation and PMF as a function of βJ for the two configurations of the dipoles.

Figure N.3(a) shows the MI for this system. This is defined by

$$I^{(2)}(X_1, X_2) = \sum_{x_1, x_2} P(x_1, x_2) \log g(x_1, x_2). \quad \text{(N.27)}$$

The corresponding average PMF is defined by

$$W^{(2)}(X_1, X_2) = \sum_{x_1, x_2} P(x_1, x_2)$$
$$\times \left[-k_B T \log g(x_1, x_2) \right]. \quad \text{(N.28)}$$

The important result for this system of two dipoles (or any two random variables) is that although the logarithm of the pair correlation can be either positive or negative, the MI is always positive or, equivalently, the average PMF is always negative [Figure N.3(b)].

The informational-theoretical interpretation of this result is that the amount of information with respect to X_1 contained

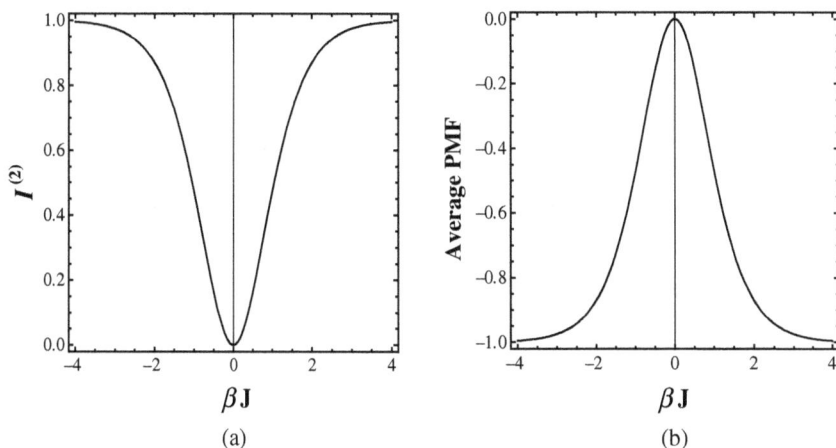

Fig. N.3. The mutual information and the average PMF as a function of βJ.

in X_2 (or vice versa) is always positive. Another equivalent interpretation is that the amount of uncertainty associated with one random variable can only decrease by knowing the second random variable.

We next turn to the case of three dipoles as in Figure 7.10. We first plot the pair MI as defined in Eq. (N.6). Note that this is defined in the system of three dipoles, and hence in Figure N.4, we denote its pair MI as $I_2^{(3)}$.

In Figure N.5, we plot first $I_3^{(3)}$ as defined in Eq. (N.4) and also $F_3^{(3)}$ as defined in Eq. (N.7). We see that $I_3^{(3)}$ is always positive. [In this case, $I_3^{(3)}$ is simply the *distance* between $P(x_1, x_2, x_3)$ and $P(x_1) P(x_2) P(x_3)$]. The quantity $F_3^{(3)}$ defined in Eq. (N.7) is positive for $\beta J > 0$ and negative for $\beta J < 0$. The latter is referred to as a frustrated system. As we have shown in Eqs. (N.20) and (N.22), $F_3^{(3)}$ [denoted F_3

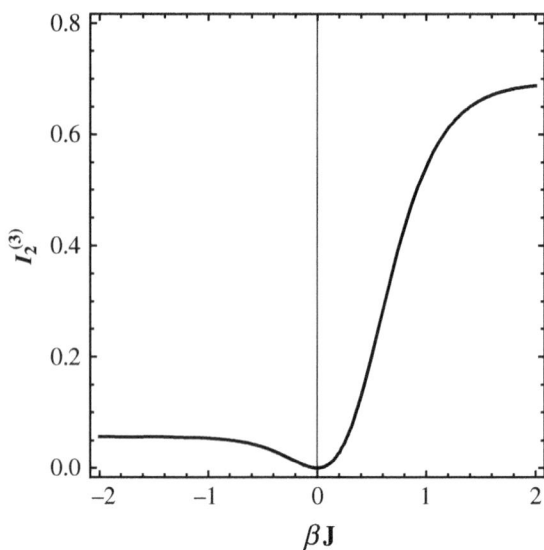

Fig. N.4. The mutual pair information in a system of three dipoles as a function of βJ.

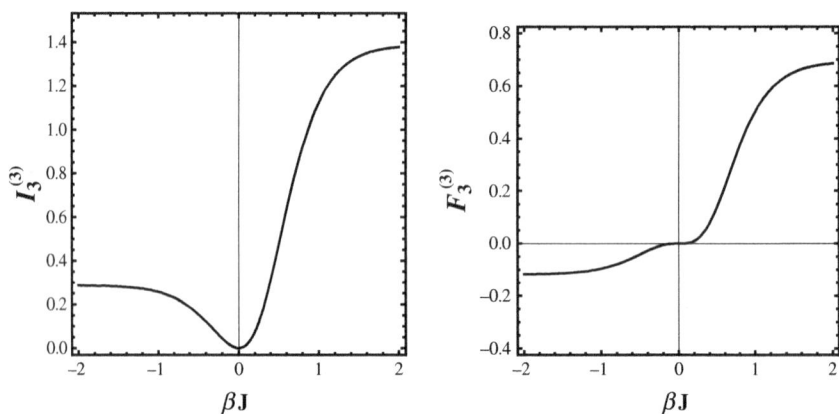

Fig. N.5. The mutual triplet information and the corresponding F function in a system of three dipoles as a function of βJ.

in Eqs. (N.20) and (N.22)] is a measure of the nonadditivity of the PMF.

The informational-theoretical interpretation of the negative sign of $F_3^{(3)}$ can be given based on Eq. (N.7). Since both $I(X_1, X_2)$ and $I(X_1, X_2 | X_3)$ must be positive quantities, it follows that $F_3^{(3)} < 0$ means that

$$I(X_1, X_2) < I(X_1, X_2 | X_3), \qquad \text{(N.29)}$$

which means that the amount of information on X_1 contained in X_2 increases when we know X_3. Another interpretation can be obtained from the definition of F_3 [see Matsuda, Eq. (2.11)],

$$F_3^{(3)}(X_1, X_2, X_3) = I_2^{(3)}(X_1, X_2) + I_2^{(3)}(X_1, X_3)$$
$$- I_2^{(3)}(X_1, X_2, X_3), \qquad \text{(N.30)}$$

where $I_2^{(3)}(X_1, X_2, X_3)$ is the amount of information contained in the *joint random variables* $X_2 X_3$ on X_1. A negative value of $F_3^{(3)}(X_1, X_2, X_3)$ means that the amount of information contained in the *joint random variables* $X_2 X_3$ on X_1 is larger than the sum of the information contained in X_2 on X_1 *and* contained in X_3 on X_1.

Epilogue

This book is about myths in protein theories — myths that have long been debunked and yet continue to flourish and propagate. In this epilogue, I will tell you about my role in the story of these myths, providing a glimpse into the survival of the (un)fittest theories, paradigms and dogmas. In doing so, I hope that you will obtain your own answer to the question that I have posed many times in the pages of this book: How can one explain the survival and the lingering of a theory despite the overwhelming evidence which disproves it? I will not answer this question directly; I believe that every experienced scientist knows the answer. My answer is "hidden" in the following stories.

In the major part of this epilogue, I will discuss five well-established myths discussed in Chapters 2–6. At the end, I will also briefly comment on some of the topics discussed in Chapter 7.

(1) The association of the $H\phi O$ effect (either solvation or interactions) with structural changes in the solvent probably originated in Frank and Evans' 1945 article. Shortly after finishing my PhD, I questioned this relationship between

the solvation Gibbs energy of simple nonpolar solutes, and the structural changes in the water induced by these solutes. In the early 1970s, I examined the relationship by using a simple mixture model approach to liquid water. This was summarized in my 1974 book. Admittedly, the mixture model approach to water, though very popular, was not accepted as a rigorous theory of water. People continued to question the validity of this approach in spite of the fact that it was firmly established, using concepts of statistical mechanics [Ben-Naim (1974)]. A more general treatment of the Gibbs energy of solvation [Ben-Naim (1978)], and of $H\phi O$ interactions [Ben-Naim (1975)], were also published. These works and their conclusions were criticized by Marcelja *et al.* (1977) and by Nemethy *et al.* (1981), respectively, claiming that my conclusions *conflict* with the second law of thermodynamics. I remember that at one of the Gordon conferences Henry Frank approached me and said that he had reviewed an article of mine on this subject. He told me that he liked the article and recommended its publication, but he also admitted that he was not sure that my proof, which was based on the mixture model approach, had a general validity. This comment encouraged me to seek for a more general argument independent of any model for water (see Chapter 2).

In 1988, I read an article by Yu and Karplus (1988) titled "A Thermodynamic Analysis of Solvation." I was both satisfied and disappointed. Although it was gratifying that they had supported my earlier publications, I found it surprising that they presented exactly the same proof that

I had presented ten years earlier, without even mentioning my work. What was even more surprising was their conclusion:

> . . . *that the solvation energy and entropy do have contributions from the change in solvent structure that are non-zero at infinite dilution. However, these contributions cancel exactly in the solvation free energy.*

This is the very same conclusion I reached several times for both solvation and $H\phi O$ interactions, and yet these authors completely ignored my earlier publications. I wrote to the senior author, and he admitted that he was not aware of my publication, and promised to write an apology in a future publication. His promise, written in the sand, was quickly swept away as the waves came rushing in. I never read an apology. The fact is that, since 1988, I have seen many authors cite the above quotation, giving credit to Yu and Karplus, while ignoring my almost-ten-years-earlier publications, where I reached exactly the same conclusions.

What is even more startling is that, in spite of my conclusions, and in spite of Yu and Karplus reproving my conclusion, there are still others, and there are many of them, who cling to the old myth that structural changes induced in the solvent *explain* the hydrophobic effects (both solvation and interactions). I fail to understand this phenomenon. The reader is invited to suggest an explanation.

(2) In 1991, while I was examining the whole question of the dominance of the $H\phi O$ interactions in protein folding, I stumbled upon a few papers by Alan Fersht where the so-called HB inventory argument is discussed. As I have discussed in Chapter 3, the origin of this argument can be

traced back to John Schellman's 1955 paper. In fact, the conclusion from Schellman's experiment was one of the main motivations for the evolution of the idea of the $H\phi O$ role in protein stability. In a paper I published in 1991, I showed that the HB inventory argument is fundamentally wrong. My arguments are discussed in Chapter 3 of this book. I sent a copy of the preprint to several of my friends, as well as to Fersht, for comments. My paper's conclusion, aside from proving the invalidity of the HB inventory argument, was that *direct* HBs do contribute significantly to protein stability. In 2007, when I visited Eugene, Oregon, I discussed this matter with Schellman. He denied that he had ever reached the conclusion attributed to him by Walter Kauzmann.

Through the years, I have read several publications in which similar conclusions were reached. It was yet another surprise to me when, in 1999, Fersht published a book where he reiterated the same "argument" which I had refuted many years earlier, completely ignoring my paper, published in 1991, and a preprint of it which was sent to him.

Can the reader please explain this? I think the answer is obvious.

Moreover, I have read many statements in recent biochemistry textbooks repeating the same old, and erroneous, conclusions that direct HB cannot contribute significantly to the stability of proteins.

I would like to add yet another anecdote in connection with my 1991 paper, which I believe is very typical and instructive.

Between the years 2010 and 2012, I submitted an article titled "Levinthal's Question Revisited, and Answered" to the *Journal of Biomolecular Structure and Dynamics*. It was sent to numerous reviewers. One of the last referees, who recommended my paper for publication, requested me to provide in my article references to three articles written by A, B and C, whose conclusions about the role of HBs in protein folding were similar to mine. The referee thought that I should give the authors credit for their idea that HBs contribute significantly to protein stability. In view of this comment, I requested the editor to send my article to other referees as I adamantly refused to provide references to those articles: first, because the role of the HB was not the main point of my article, second, and more importantly, although all three articles (A, B and C) discussed the role of the HB, they reached the same conclusion which I had made 20 years earlier in my 1991 article. I felt that it would be ironic to acknowledge their papers when these authors failed to acknowledge the same conclusion which I had reached and published much earlier. After requesting the editor to send my article yet again, but to an unbiased referee, I finally got a review from the last referee. The referee wrote that he had read my 1991 article, as well as the three articles by A, B and C (to which I was required to provide references). The referee concluded, aside from recommending that my paper be published, that the author did not have to provide references to A, B and C, but *must* provide references to his own article written in 1991. I did as requested, and that ended the saga of this new article.

(3) I was always fascinated by the hydrophobic effect. In the early 1970s, I wrote a review article on what was known at that time about the various $H\phi O$ effects. I noticed that there

was no single measure of what I referred to as "a pure measure of $H\phi O$ interactions." So I created one. I published this in several articles in *The Journal of Chemical Physics*. Most of those articles were reviewed by Kauzmann, who consistently sent me copies of his review by mail (e-mail did not exist yet).

The main idea underlying the design of the new measure of $H\phi O$ interactions was to separate the direct from the indirect interactions between two nonpolar solutes. This separation immediately suggested to me that, from the viewpoint of the solvent molecules, the direct solute–solute interaction is irrelevant to the study of $H\phi O$ interactions. Therefore, the solvation Gibbs energy of any hydrocarbon molecule can be viewed as the solvation Gibbs energies of two or more smaller hydrocarbon molecules at very short distances between them.

The new measure was used to characterize the $H\phi O$ interaction between two small hydrocarbon molecules in different solvents. It soon became clear that water as a solvent exhibits an unusually strong $H\phi O$ interaction compared to all other solvents for which the relevant data were available. I summarized all that was known at that time about $H\phi O$ solvation and $H\phi O$ interactions in a monograph published in 1980.

Although the new measure of $H\phi O$ interactions established the uniqueness of water as a solvent for small hydrocarbon molecules, there was no evidence that these $H\phi O$ interactions, or $H\phi O$ effects (the more general term), are important in stabilizing protein structure. I have discussed this matter several times with Kauzmann. He agreed that there is no direct evidence for the contention about the dominance of the $H\phi O$ effects in proteins but, having no other factor in sight which could explain the stability of proteins, he believed that his own conjecture regarding the role of the $H\phi O$ effects would be corroborated.

I did not share that belief, however, and in the preface to my 1980 book, I expressed my doubts regarding the role of $H\phi O$ effects in protein folding and protein–protein association.

Notwithstanding my doubts regarding the role of the $H\phi O$ effects in proteins, I believed that these effects are quite unique to liquid water, and as such deserve systematic study. I was amazed, however, to see so many scientists present Kauzmann's conjecture as the "most important," the "dominant force," etc. in protein folding. Such unwarranted statements pervaded the literature of biochemistry. Authors of textbooks and research scientists referred to the "well-known" and "well-established" fact of the dominant role of $H\phi O$ effects in protein folding without ever questioning its validity.

A turning point in the late 1980s occurred, while I was spending some time at the NIH. I undertook the ambitious task of establishing an *inventory* of all possible solvent-induced effects in protein folding and protein–protein association. This analysis led to two important conclusions which had profound implications regarding the role of $H\phi O$ effects in protein folding. First, it was found that neither the $H\phi O$ solvation nor the interaction between two $H\phi O$ molecules feature in the Gibbs energy of the process of protein folding. Instead, the *conditional* $H\phi O$ solvation and the *conditional* $H\phi O$ interactions are operative in the folding process. It was also found that the magnitude of each of the *conditional* $H\phi O$ effects was much smaller than that of the corresponding $H\phi O$

effect. This finding, in itself, was already indicative of the not-so-important role of the $H\phi O$ effects in protein folding.

Second, in addition to the well-told and well-recognized $H\phi O$ effects, it was found that some of the $H\phi I$ effects are far stronger than the corresponding $H\phi O$ effects. These findings had shifted my focus from the $H\phi O$ effects to the $H\phi I$ effects. The submitted articles on these findings were bombarded with vigorous objections by referees. Needless to say, I found most by the criticisms of the reviewers either irrelevant or insensible. One referee, who recommended the rejection of my article, had only one critical comment:

> *The main problem with the article is that it is written in a tone of a Nobel laureate, whose ideas have been generally accepted by the world. The author has a lot of convincing to do before his ideas gain wide acceptance.*

This was not the only case of a rejection based on the "tone" of my writings. How could have I convinced the biochemical community, being denied the opportunity to publish my views? Of course, such comments and many others did not deter me from publishing my views. I went on to publish my views in a series of articles, and summarized them, culminating with the publication of my book in 1992, and more recently in 2011d.

In spite of the overwhelming evidence in favor of the $H\phi I$ compared to the $H\phi O$ effects, people continue to cling to the $H\phi O$ dogma. Can the reader please explain this phenomenon? Some scientists not only ignore the evidence in favor of the $H\phi I$ effects, but also actively discourage others from even questioning the "dominance of the $H\phi O$

effects." Recently, I received several e-mails from authors who gave references to my articles and challenged the mainstream dogma. They were required by referees to remove any statement challenging the $H\phi O$ dogma.

Finally, and even more recently, I sent a preprint to a friend for comments. He wrote back saying that he fully agreed with me, but begged me, "Please do not mention my name in your acknowledgement." I was appalled by his request. As everyone knows, mentioning the name of someone in an acknowledgement does not imply that the person *agrees* with what is written in the article.

(4) In 2012, I published two articles titled "Levinthal's Paradox Revisited and Dismissed" and "Levinthal's Question Revisited and Answered." It is very clear from the titles alone that in these articles I have made a clear distinction between what has been referred to in the literature as the "Levinthal paradox" and as the "Levinthal question" raised in Levinthal's article. This distinction is discussed in great detail in these two articles.

After the publication of the second article in *JBSD*, the editor sent a copy of it to about 40 scientists for their comments. According to the editor, about 20 responded to his request to write some kind of an open review of this article.

A collection of the reviews, along with my responses, was published in *JBSD* in the same year, 2012. Most of the comments were favorable and I enjoyed reading them. Some were funny and clearly revealed that the authors either did not understand or perhaps did not even read what I wrote.

Here are two examples:

Thus, unfortunately, Ben-Naim (2012) becomes another addition to the vast body of literature that has failed to quantitatively address Levinthal's question in a manner even remotely close to the elegant calculation known as Levinthal's paradox.

— Aditya Mittal and Chanchal Acharya

This statement is quite unfortunate. My article has nothing to do with Levinthal's paradox. This (non)paradox is discussed separately in another article [Ben-Naim (2012b)] and in a recent monograph [Ben-Naim (2013)]. This is precisely the reason why I titled my article with "Levinthal's question" and not "Levinthal's paradox."

The distinction between the two was made clear in my article, which was unfortunately missed.

Alexei V. Finkelstein and Sergiy O. Garbuzynskiy wrote a response with the following title "Levinthal's Question Answered... Again?"

My answer is: no, not again! To the best of my knowledge, the answer I have given to Levinthal's question was *never* given before. The authors found *"nothing non-trivial..."* in my article. Unfortunately, what they have written shows clearly that they misunderstood all my "trivial" points. To give only one example (out of the many meaningless statements made in this article), the authors conclude:

Summarizing, I have to say that I cannot find in the paper by Ben-Naim anything more than a complicated reincarnation of the old good funnel model of protein folding.

My paper *criticizes* the funnel model as an *unfounded* model and is not a *"reincarnation* of the old good funnel model." It is strange that the authors read in my article *exactly the opposite* of what I wrote!

To conclude, my contribution to the Levinthal paradox was not to *solve* the paradox, but rather to *dismiss* it. Instead, I focused on a very important question raised by Levinthal: What are the factors which *speed* and *guide* the folding of proteins? My immediate answer to this question was straightforward. There must be strong *forces* that *speed* and *guide* the folding of proteins to their 3D native structure. This answer raised a new question: What are the strong forces? And my answer is: Most likely the solvent-induced forces exerted on $H\phi I$ groups along the chain of amino acids.

This answer is very different from the published view that the sequence of amino acids has evolved to obey the principle of minimum frustration, and that the funnel-like shape of the GEL explains both the speed and the guidance of the folding process.

(5) The PFP is relatively new to me. I first learned about it from Harry Saroff about 20 years ago. He told me, among many other things, that Anfinsen was lucky to choose a well-behaved protein for his experiments which led him to pronounce the thermodynamic hypothesis. Upon hearing about this hypothesis for the first time, I had the feeling that it is nothing but a statement of the second law of thermodynamics applied to a system at constant T, P, N.

It was only a few years ago that I realized that Anfinsen's hypothesis can be interpreted in two ways. One, as a statement of the second law, i.e. that the Gibbs energy of the entire system under constant T, P, N at equilibrium must be at a minimum. The second way is that the 3D native structure of the protein must be at the global minimum of the GEL. Both of these interpretations can be "extracted" from Anfinsen's hypothesis (see the quotation at the beginning of Chapter 6). Both of them seem to follow from the second law. Therefore, both of these interpretations seem to be equivalent and true. Yet, intuitively, I felt that something in that conclusion seemed to be amiss. I could not spell out what was bothering me. After a long period of incubation, the picture became sharper and clearer: the two interpretations are not equal — one is trivial and the other is wrong.

This finding was quite unexpected for me. Looking through the literature, I was shocked to learn that most people who work in the field have actually adopted the *wrong interpretation* of Anfinsen's hypothesis. People have spent a great deal of effort in developing sophisticated algorithms to locate the global minimum in the EL (or the PES or the GEL), in the hope that this will lead to the discovery of the "folding code," or the "prediction" of the protein structure from the sequence.

It took me a few more months and a great deal of writing and rewriting, honing and pinpointing the reasons for this almost colossal misinterpretation of Anfinsen's hypothesis. Immersing myself in textbooks on thermodynamics revealed to me the culprit behind such a gross misinterpretation. Most textbooks teach us the second law in terms of *maximum entropy*, or minimum Helmholtz or Gibbs energy. Most textbooks also provide a few examples of maximum entropy or minimum Gibbs energy (see the first two examples in Appendix J). Almost none of the textbooks spell out the *variable* with respect to which the Gibbs energy has a minimum. (For details, see Chapter 6 and Appendix K.)

This gap in the teaching of the second law is the main culprit in the misinterpretation of Anfinsen's hypothesis. More specifically, the Gibbs energy of the protein in solution is viewed as a function of the *conformation*, i.e. of the coordinates specifying the conformation — $G(T, P, N; \phi_1, \ldots, \phi_m)$. Anfinsen said that the Gibbs energy of the native structure is at a minimum. This was translated as the minimum of the Gibbs energy with respect to the variable $\phi_1, \ldots, \phi_m$. In modern language, Anfinsen's hypothesis was interpreted in terms of a global minimum in the GEL (or, worse, in the EL or the PES; see Chapter 6). That was a serious pitfall of Anfinsen's hypothesis. This is also the title of an article that I sent to *Chemical Physics Letters* (2011a).

Two years later, I wrote another article, entitled "Myths and Verities in Protein Folding Theories, Part I: Anfinsen's Hypothesis and the Search for the Global Minimum in the Gibbs Energy Landscape." I thought that this paper would be an important one which could save a lot of people a lot of effort, as well as time and money.

I sent the manuscript to *The Journal of Chemical Physics*, and after a month I got a beautiful review from a referee who recommended the publication of the article subject to a few *minor* and *optional* corrections. I was glad to receive the letter of the associate editor:

Dear Professor Ben-Naim,

I have just received the review of your manuscript from the referee(s) to whom it was sent. The comments are enclosed.

When you return the double-spaced revised manuscript to us, along with a letter describing the revisions, it will be accepted for publication in The Journal of Chemical Physics.

You may submit a revised version (double-spaced and in preprint format) and a letter explaining the revisions at:

Sincerely yours,

Murugappan Muthukumar

Associate Editor

The Journal of Chemical Physics

I have received such letters in the past, when a referee recommended publication and the paper was accepted, pending some optional changes.

As soon as I received the acceptance letter, I sent about ten preprints to friends and colleagues who are interested in this field. Most of the responses were favorable. Two people did not like the article, and one asked me to which journal I had sent the article.

I made a few (minor) corrections to the article and resubmitted it to the editor.

A couple of months had passed and there was still no sign of the proofs. I sent an inquiry to the editor, and got a short note from the secretary of *JCP* instead, informing me that the article was still being considered by the referees for publication. That came as a complete shock to me, as I had never heard of a case where a paper which had already been *accepted for publication* had to go through the rigmarole of further consideration. I suspected something fishy regarding the handling of my article. My suspicion was confirmed when the associate editor approached me at a conference in Albany and *apologized* for what had happened. Although he did not explain to me what had happened, the fact that he saw the need to *apologize* made it very clear to me. At that moment, I understood that I had made a mistake in sending preprints to a few colleagues whom I quoted in the article.

Soon after my encounter with *JCP*'s associate editor, I got two more reviews. One recommended publication after some correction, while the other rejected the article. It was clear to me that the second referee had no idea what the paper was all about. His comments were between insensible and irrelevant, or both. Here is an example:

> *The author makes some points that are true, albeit well known to practitioners.... The author is making arguments that are misleading, or in many cases well-known.*

What I wrote was anything but "well known." What I wrote was about the "well-known" *erroneous* association of the structure of proteins with the global minimum of the GEL. If any reader can prove to me that the content of Chapter 6 is "well-known," I promise to remove it from the book.

I will leave it to the reader to reach his or her own conclusion from this story.

The article was soon published in another journal, and it is reproduced with more details in Chapter 6 of this book. I still owe the first referee an acknowledgement of his or her understanding of the article, as well as his or her recommendation.

I had originally planned on writing only six chapters dealing with the five "well-known" and "well-established" dogmas. However, after writing these chapters I decided to add Chapter 7, where I have discussed some would-be, or could-be, new dogmas.

As I stated in the preface, the PFP is not an easy problem. However, the literature presents this problem in monstrous and exaggerated terms. I believe that the PFP became bigger than what it really is mainly because of searching in the wrong directions, studying the $H\phi O$ effects instead of the full spectrum of solvent-induced effects — studying the entire shape of the GEL instead of focusing only on the tiny region under the GEL which is relevant to proteins under physiological conditions. I should also add that most of what people know (or claim to know) about the GEL of proteins is obtained from studies of lattice models for proteins. In my opinion, lattice models are totally irrelevant to the PFP, and much of the effort in their study will sooner or later be relegated to the shadows of obsolescence.

One final comment about the use of the term "principle" in connection with protein folding. In Chapter 7, I have discussed two: the "principle of consistency" and the "principle of minimal frustration." I have no idea what these principles are. The literature is replete with statements which *describe* the principles sometimes in vague terms, and sometimes they are made equivalent, but nowhere could I find a simple and clear explanation of them. Whatever these principles stand for, one thing is very clear: they are anything but *principles*.

References

Amano, K. I., Suzuki, K., Fukuma, T., *et al.* (2013), *J. Chem. Phys.* **139**, 224710.

Amit, D. J. (1989), *Modelling Brain Function: The World of Attractor Neural Networks*. Cambridge University Press, New York.

Anfinsen, C. B. (1973), Principles that govern the folding of protein chains, *Science*, New Series **181**, 223–230.

Baldwin, R. L. (1994), Matching speed and stability, *Nature* **369**, 183.

Baldwin, R. L. (1995), The nature of protein folding pathways: the classical versus the new view, *Proc. Natl. Acad. Sci. USA* **47**, 1309.

Ben-Naim, A. (1970), *Trans. Faraday Soc.* **66**, 2749.

Ben-Naim, A. (1974), *Water and Aqueous Solutions: Introduction to a Molecular Theory*, Plenum Press, New York.

Ben-Naim, A. (1975a), Hydrophobic interactions and structural changes in the solvent, *Biopolymers* **14**, 1337.

Ben-Naim, A. (1978), *J. Phys. Chem.* **82**, 874.

Ben-Naim, A. (1980), *Hydrophobic Interactions*, Plenum Press, New York.

Ben-Naim, A. (1989), Solvent-induced interactions: hydrophobic and hydrophilic phenomena, *J. Chem. Phys.* **90**, 7412–7525.

Ben-Naim, A. (1990a), Solvent-induced forces in protein folding, *J. Phys. Chem.* **94**, 6893–6895.

Ben-Naim, A. (1990b), Solvent-effects on protein association and protein folding, *Biopolymers* **29**, 567.

Ben-Naim, A. (1991a), Strong forces between hydrophilic macro-molecules; implications in biological systems, *J. Chem. Phys.* **93**, 8196–8210.

Ben-Naim, A. (1991b), The role of hydrogen bonds in protein folding and protein–protein association, *J. Phys. Chem.* **95**, 1437.

Ben-Naim, A. (1992), *Statistical Thermodynamics for Chemists and Biochemists*, Plenum Press, New York.

Ben-Naim, A. (2001), *Cooperativity and Regulation in Biochemical Systems*, Plenum Press, Kluwer, New York.

Ben-Naim (2006), *Molecular Theory of Solutions*, Oxford University Press, Oxford.

Ben-Naim, A. (2008), *A Farewell to Entropy: Statistical Mechanics Based on Information*, World Scientific, Singapore.

Ben-Naim, A. (2009), *Molecular Theory of Water and Aqueous Solutions, Part I: Understanding Water*, World Scientific, Singapore.

Ben-Naim, A. (2011a), *Chem. Phys. Lett.* **511**, 126.

Ben-Naim, A. (2011b), *J. Chem. Phys.* **135**, 085104.

Ben-Naim, A. (2011c), Pitfalls in Anfinsen's thermodynamic hypothesis, *Chem. Phys. Lett.* **511**, 1–3, pp. 126–128. doi:10.1016/j.eplett.2011.05.049.

Ben-Naim, A. (2011d), *Molecular Theory of Water and Aqueous Solutions, Part II: The Role of Water in Protein Folding, Self-Assembly and Molecular Recognition*, World Scientific, Singapore.

Ben-Naim, A. (2012a), Some no longer unknown of science, *Open J. Biophys.* **2**, 9–11.

Ben-Naim, A. (2012b), Levinthal's paradox revisited and dismissed, *Open J. Biophys.* **2**(2), 22–32. doi:10.4236/ojbiphy.2012.2204.

Ben-Naim, A. (2012c), Levinthal's question revisited and answered," *J. Biomol. Struct. Dynam.* **30**(1), pp. 113–124. doi.10.1080/07391102.2012.674286.

Ben-Naim, A. (2012d), *Entropy and the Second Law: Interpretation and Miss-interpretations*, World Scientific, Singapore.

Ben-Naim, A. (2013), *The Protein Folding Problem and Its Solutions*, World Scientific, Singapore.

Ben-Naim, A. (2015), *Information, Entropy, Life, and the Universe*, World Scientific, Singapore.

Ben-Naim, A. and Friedman, H. L. (1967), *J. Phys. Chem.* **71**, 448.

Birshthein, T. M. and Ptitsyn, O. B. (1966), *Conformations of Macromolecules*, Wiley-Interscience, New York.

Bolen, D. W. and Rose, G. D. (2008), *Annu. Rev. Biochem.* **77**, 339–362.

Bryngelson, J. D., Onuchic, J. N., Socci, N. D. and Wolynes, P. G. (1995), *Proteins. Struct. Funct. Genet.* **21**, 167.

Bryngelson, J. D. and Wolynes, P. G. (1987), *Proc. Natl. Acad. Sci. USA* **84**, 7524.

Bryngelson, E. I. and Gutin, A. M. (1989), *Biophys. Chem.* **34**, 187.

Busch, S., Bruce, C. D., Redfield, C., *et al.* (2013), *Angew. Chem. Ed.* **52**, 13091.

Chiti, F. and Dobson, C. M. (2006), *Annu. Rev. Biochem.* **75**, 333–366.

Cordes, M. H. J., Davidson, A. R. and Sauer, R. T. (1996), *Curr. Opin. Struct. Biol.* **6**, 3.

Creighton, T. E. (1993), *Proteins, Structure and Molecular Properties*, W. H. Freeman and Company, New York.

Dawkins, R. (1987), *The Blind Watchmaker*, Norton, New York.

Delvin, M. T. (2006), *Textbook of Biochemistry with Clinical Correlations*, 6th edn., Wiley-Liss, Hoboken, New Jersey.

Dias, C. L. *et al.* (2011), *J. Chem. Phys.* **134**, 065106.

Dias, C. L. and Chan, H. S. (2014), *J. Phys. Chem.* **118**, 7488.

Dill, K. A. (1985), Theory of folding and stability of globular proteins, *Biochemisty* **24**, 1501.

Dill, K. A. (1990), Dominant forces in protein folding, *Biochemistry* **29**, 7133.

Dill, K. A., Fiebig, K. M. and Chan, H. S. (1993), *PNAS* **90**, 1942.

Dill, K. A. and Chan, H. S. (1997), *Nat. Struct. Biol.* **4**, 10.

Dill, K. A. (1999), Polymer principle and protein folding, *Protein Sci.* **8**, 1–166.

Dill, K. A., Ozcan, S. B., Shell, M. S. and Weikl, T. R. (2008), *Annu. Rev. Biophys.* **37**, 289.

Dinner, A. R., Sali, A., Smith, L. J., *et al.* (2000), *Trends Biochem. Sci.* **25**, 331–339.

Dobson, C. M. and Karplus, M. (1999), *Curr. Opin. Struct. Biol.* **9**, 91–101.

Durell, S., Brooks, B. R. and Ben-Naim, A. (1994), Solvent induced forces between two hydrophilic groups, *J. Phys. Chem.* **98**, 2198–2202.

England, J. L. and Haran, G. (2010), *PNAS* **107**, 14519.

Englander, S. W. and Mayne, L. (2014), *PNAS* USA, **111**, 15873.

Eisenberg, D. and Kauzmann, W. (1969), *The Structure and Properties of Water*, Oxford University Press, Oxford.

Eley, D. D. (1939), *Trans. Faraday Soc.* **35**, 1281.

Eley, D. D. (1944), *Trans. Faraday Soc.* **40**, 184.

Fang, Y. (2013) Ben-Naim's "pitfall": Don Quixote's windmill, *Open J. Biophys.* **3**, 13–21.

Fang, Y. (2015), Why Ben-Naim's deepest pitfall does not exist, *Open J. Biophys.* **5**, 45.

Fasman, G. D. (1989), *Prediction of Protein Structure and the Principles of Protein Conformation*, Plenum Press, New York.

Ferreiro, D. U., Hegler, J. A., Komives, E. A. and Wolynes, P. G. (2007), *PNAS* **104**, 19819.

Ferreiro, D. U., Hegler, J. A., Komives, E. A. and Wolynes, P. G. (2011), *PNAS* **108**, 3499.

Ferreiro, D. U., Komives, E. A. and Wolynes, P. G. (2013), *Frustration in Biomolecules*, arXiv: 1312.0867v1[q-bio.BM].

Fersht, A. (1985), *Trends Biochem. Sci.* **9**, 145.

Fersht, A. (1987), The hydrogen bond in molecular recognition, *TIBS* **12**, 301.

Fersht, A. (1999), *Structure and Mechanism in Protein Science: A Guide to Enzyme Catalysis and Protein Folding*, W. H. Freeman, New York.

Finkelstein, A. V. (1997), *Curr. Opin. Struct. Biol.* **7**, 60–71.

Finkelstein, A. V. and Galzitskaya, O. V. (2004), Physics of protein folding, *Phys. Life Rev.* **1**, 23–56.

Fischer, K. H. and Hertz, J. A. (1991), *Spin Glasses*, Cambridge University Press, New York.

Fowler, R. and Guggenheim, E. A. (1956), *Statistical Thermodynamics*, Cambridge University Press.

Franks, H. S. and Evans, M. W. (1945), *J. Chem. Phys.* **13**, 507–532.

Frauenfelder, H., Sligar, S. G. and Wolynes, P. G. (1991), *Science* **254**, 1598.

Gō, N. (1983), *Annu. Rev. Bioeng.* **12**, 183.

Gō, N. (1984), *Annu. Rev. Bioeng.* **18**, 149.

Goldberg, M. E. (1985), *TIBS* **10**, 388.

Goldstein, M., Fredj, E. and Gerber, R. B. (2011), *J. Comput. Chem.* **32**, 1785.

Goldtzvik, Y., Goldstein, M. and Gerber, R. B. (2013), *J. Chem. Phys.* **415**, 168.

Grana-Montes, R. and Ventura, S. (2013), *J. Biomol. Struct. Dynam.* **31**, 970.

Graziano, G. (2013), *J. Phys. Chem. B.* **117**, 2153.

Haber, E. and Anfinsen, C. B. (1961), Studies on the reduction on reformation of protein disulfide bonds, *J. Biol. Chem.* **236**, 1361–1363.

Haberfield, P., Kivuls, J., Haddad, M. and Rizzo, T. (1984), Enthalpies, free energies, and entropies of transfer of phenols from nonpolar solvents to water, *J. Phys. Chem.* **88**, 1913.

Hecht, M. H., Das, A., Go, A., *et al.* (2004), *Protein Sci.* **13**, 1711.

Helling, R., Li, H., Melin, R., *et al.* (2001), *J. Mol. Graph. Model.* **19**, 157–167.

Hill, T. L. (1956), *Statistical Mechanics Principles and Selected Applications*, Addison-Wesley, Reading, Massachusetts.

Honig, B. (1999), *J. Mol. Biol.* **293**, 283–293.

Jenik, M., Parra, R. G., Radusky, L. G., *et al.* (2012), *Nucl. Acid Res.* **40**, W348.

Kamketar, S., Schiffer, J. M., Xiong, H., *et al.* (1993), *Science* **262**, 1680.

Karplus, M. (1997), *Fold. Des.* **2**, S69–75.

Karplus, M. (2011), Behind the folding funnel diagrams, *Nat. Chem. Biol.* **7**, 401–404.

Kauzmann, W. (1959), Some factors in the interpretation of protein denaturation, *Adv. Protein Chem.* **14**; *Protein Sci.* **2**, 671.

Kauzmann, K. (1993), *Protein Sci.* **2**, 671.

Kennedy, D. and Norman, C. (2005), What don't we know? *Science* **309**, 75.

Kessel, A. and Ben-Tal, N. (2011), *Introduction to Proteins: Structure, Function and Motion*, CRC Press, Taylor and Francis Group, New York.

Kim, D. E., Gu, H. and Baker, D. (1998), *Proc. Natl. Acad. Sci. USA* **95**, 4982.

King, N. P., Yeates, E. O. and Yeates, T. O. (2007), *J. Mol. Biol.* **373**, 153.

Kirkwood, J. G. (1935), *J. Chem. Phys.* **3**, 300.

Kirkwood, J. G. and Buff, F. P. (1951), *J. Chem. Phys.* **19**, 774.

Kirkwood, J. G. (1954), in *Symposium on the Mechanism of Enzyme Action*, eds. W. D. McElroy and B. Glass, Johns Hopkins University Press, Baltimore, Maryland.

Kolata, G. (1986), *Science* **233**, 1037.

Leach, A. (2001), *Molecular Modelling: Principles and Applications*, 2nd edn., Prentice Hall, UK.

Lee, B. (1991), *Biopolymers* **31**, 993.

Levinthal, C. (1968), Are there pathways for protein folding, *J. Chem. Phys.* **65**, 44.

Levinthal, C. (1969), in *Mossbauer Spectroscopy in Biological Systems*, Proceedings of a meeting held at Allerton House, Monticello, Illinois, eds. J. T. P. De Brunner and E. Munck, University of Illinois Press, pp. 22–24.

Levinthal, C. (1969), How to fold graciously, *Mössbauer Spectroscopy in Biological Systems Proceedings* **67**(41), 22–26.

Levy, Y., Cho, S. S., Shen, T., *et al.* (2005), *PNAS* **102**, 2373.

Li, Z. and Scheraga, H. A. (1987), *Proc. Natl. Acad. Sci. USA* **84**, 6611.

Liwo, A., Lee, J., Ripoll, D. R., *et al.* (1999), *Proc. Natl. Acad. Sci. USA* **96**, 5482.

Lloyd, S. (2006), *Programming the Universe*, Alfred A. Knopf, New York.

Mallamace, F., Corsaro, C., Malamace, D., *et al.* (2014), *Comput. Struct. Biotechnol. J.*

Marcelja, S., Mitchell, D. J., Ninham, B. W. and Sculley, M. J. (1977), *J. Chem. Soc. Faraday Trans.* **73**, 630.

Matsuda, H. (2000), *Phys. Rev. E* **62**, 3096.

Matthews, C. R. (1993), Pathways of protein folding, *Annu. Rev. Biochem.* **62**, 653.

Meeron, E. (1957), *J. Chem. Phys.* **27**, 1238.

Mezei, M. and Ben-Naim, A. (1990), Calculation of the potential of mean force between water molecules in fixed relative orientation in liquid water, *J. Chem. Phys.* **92**, 1359–1361.

Morcos F., Schafer, N. P., Cheng, R. R., *et al.* (2014), *PNAS* **111**, 12408.

Münster, A. (1974), *Statistical Thermodynamics*, Vol. II, Springer-Verlag, Berlin.

Myers, J. K. and Pace, C. N. (1996), *Biophys. J.* **71**, 2033.

Nemethy, G., Peer, W. J. and Scheraga, H. A. (1981), *Annu. Rev. Biophys. Bioeng.*, **10**, 459.

Noel, J. K., Sulkowska, J. I. and Onuchic, J. N. (2010), *PNAS* **107**, 15403.

Nolting, B. (2006), *Protein Folding Kinetics: Biophysical Methods*, 2nd edn., Springer-Verlag, Germany.

Onuchic, J. N. and Wolynes, P. G. (2004), *Curr. Opin. Struct. Biol.* **14**, 70.

Orevi, T., Rahamin, G., Shemesh, S., *et al.* (2014), *BAMS* **10**, 169.

Pace, C. N. (2009), *Nat. Struct. Mol. Biol.* **16**, 681–682.

Pace, C. N., Scholtz, J. M. and Grimsley, G. R. (2014), *FEBS Lett.* **588**, 2177.

Papoulis, A. and Pillain, S. U. (2002), *Probability Random Variables and Stochastic Processes*, McGraw-Hill, Boston.

Pauling, L. (1939), *The Nature of the Chemical Bond*, 1st edn., Cornell University Press, Ithaca, New York.

Pauling, L. (1948), *The Nature of the Chemical Bond*, 2nd edn., Cornell University Press, Ithaca, New York.

Pauling, L. (1960), *The Nature of the Chemical Bond*, 3rd edn., Cornell University Press, Ithaca, New York.

Pierotti, R. A. (1963), *J. Phys. Chem.* **67**, 1840.

Pierotti, R. A. (1965), *J. Phys. Chem.* **69**, 281.

Pierotti, R. A. (1967), *J. Phys. Chem.* **71**, 2366.

Plotkin S. S. and Onuchic, J. N. (2002), *Quart. Rev. Biophys.* **35**, 111.

Prabhu, N. and Sharp, K. (2006), *Chem. Rev.* **106**, 1616.

Prentiss, M. C., Hardin, C., Eastwood, M. P., *et al.* (2006), *J. Chem. Theory Comput.* **2**, 705.

Privalov, P. L. (1990), Cold denaturation of proteins, *Crit. Rev. Biochem. Mol. Biol.* **25**(4), 281–305.

Rice, S. A. and Lekner, J. (1965), *J. Chem. Phys.* **42**, 3559.

Rice, S. A. and Young, D. A. (1967), *Discuss. Faraday Soc.* **43**, 16.

Rose, G. D., Fleming, P. J., Banavar, J. R. and Maritan, A. (2006), *Proc. Natl. Acad. Sci. USA* **103**, 16623.

Rosenkrantz, R. D. (1989), *Papers on Probability, Statistics and Statistical Physics*, ed. E. T. Jaynes, Kluwer, Boston.

Saltpeter, E. E. (1958), *Am. Phys. NY* **5**, 183.

Schellmann, J. A. (1955), The stability of hydrogen-bonded, peptide structures in aqueous solution, *Comp. Rend. Lab. Carlsberg. Ser. Chim.* **29**, 230–259.

Schellmann, J. A. (1955), The thermodynamics of urea solutions and the heat of formation of the peptide hydrogen bond, *Comp. Rend. Lab. Carlsberg. Ser. Chim.* **29**, 223, 230.

Schlick, T. (2002), *Modelling and Simulation: An Interdisciplinary Guide*, Springer-Verlag, USA.

Shakhnovich, E. I. and Gutin, A. M. (1989), *Biochem. Phys. Chem.* **34**, 187.

Shakhnovich, E. (2006), *Chem. Rev.* **106**, 1559.

Singer, A. (2004), *J. Chem. Phys.* **121**, 3657.

Snyder, P. W., Lockett, M. R., Moustakas, D. M. and Whitesides, G. M. (2013), *Eur. Phys. J. Spec. Top.*, EDP Sciences, Springer-Verlag.

Srinivasan, R. and Rose, G. D. (2002), *Methinks it is a folding curve*, *Biophysics. Chem.* 101–102, 167–171.

Tanford, C. and Reynolds, J. (2001), *Nature's Robots: A History of Proteins*, Oxford University Press, Oxford.

Tompa, P. and Han, K. H. (2012), *Phys. Today*, August 2012, 64.

Uversky, V. N. (2002), *Protein Sci.* **11**, 739.

Uversky, V. N. and Dunker, A. K. (2010), *Biochim. Biophys. Acta* **1804**, 1231.

Uversky, V. N. (2013), Unusual biophysics of intrinsically disordered proteins, *Biochim. Biophys. Acta* **1834**, 932–951.

Voet, D. J., Voet, J. G. and Pratt, C. W. (2008), *Principles of Biochemistry*, 3rd edn., John Wiley and Sons, Hoboken, New Jersey.

Wales, D. J. and Doye, J. P. K. (1997), *J. Phys. Chem. A* **101**, 5111.

Wales, D. J. (2003), *Energy Landscapes.* Cambridge University Press, Cambridge, UK.

Wales, D. J. (2012), *Philos. Trans. Roy. Soc. A* **370**, 2877.

Wang, H. and Ben-Naim, A. (1997), Solvation and solubility of globular proteins, *J. Phys. Chem. B* **101**, 1077–1086.

Wedemeyer, W. J. and Scheraga, H. A. (2001), *Encyclopedia of Life Sciences*, John Wiley and Sons, p. 1.

Wolfenden, R. (2007), Experimental measures of amino acid hydrophobicity, and the topology of transmembrane and globular proteins, *J. Gen. Physiol.* **129**, 357.

Wolynes, P. G. (2001), *Proc. Am. Philos.* **145**, 555.

Wolynes, P. G., Onuchic, J. N. and Thirumalai, D. (1995), *Science* **267**, 1619.

Wolynes, P. G. (2005), *Philos. Trans. Roy. Soc. A* **363**, 453.

Wu, J. and Prauznitz, J. M. (2008), *Proc. Natl. Acad. Sci. USA* **105**, 4.

Yang, A. S., Sharp, K. A. and Honig, B. (1992), *J. Mol. Biol.* **227**, 889–900.

Yang, S., Cho, S. S., Levy, *et al.* (2004), *PNAS* **101**, 13789.

Yeates, T. O., Norcross, T. S. and King, N. P. (2007), *Curr. Opin. Chem. Biol.* **11**, 595–603.

Yu, H. A. and Karplus, M. (1988), *J. Chem. Phys.* **89**, 2366.

Zwanzig, R., Szabo, A. and Bagchi, B. (1992), Levinthal's paradox, *Proc. Natl. Acad. Sci. USA* **89**, 20.

Index

.

www.ingramcontent.com/pod-product-compliance
Lightning Source LLC
Chambersburg PA
CBHW052110230326
41599CB00055B/5390